The BMC Engineering Companion

By

Tony Cripps

AF885161

Publisher Information

Copyright Fischer-Cripps Laboratories Pty Ltd, 2023

All rights reserved. No part of this publication may be reproduced, stored in a retrieval system or transmitted in any form or by any means, electronic, mechanical, photocopying, recording or otherwise, without the written prior permission of the publisher.

Author: Tony Cripps
Title: The BMC Engineering Companion

ISBN 978-0-6452067-4-6

Published by Fischer-Cripps Laboratories Pty Ltd.
PO Box 9 Forestville NSW 2087 Australia

17th Jul 2024

www.leylandaustralia.com.au

NATIONAL LIBRARY OF AUSTRALIA

A catalogue record for this book is available from the National Library of Australia

Contents

Preface

In modern times, BMC/Leyland cars are classic vehicles of a bygone era, and as such, require specialist knowledge for their upkeep. Most practitioners rely on the official workshop manuals and parts lists for repair and service information. I am of the view that although very useful, it is often very advantageous to know more about how various systems work in order to get the most from them.

This book presents what looks at first sight like the usual array of topics which make up a workshop manual, but each chapter is something a bit unusual. In each section, an engineering discussion is presented followed by some practical examples that relate to BMC/ Leyland Australia vehicles. I hope the engineering content adds another dimension to your ownership and appreciation of BMC/Leyland vehicles.

I wish you happy reading.

Tony Cripps

About the Author

Tony Cripps has a Certificate in Automotive Engineering from Sydney Technical College, a Bachelor of Applied Science from the University of Technology, Sydney, (UTS), and a PhD from the University of Sydney. He has worked as a technical officer at the ex-Leyland Australia vehicle emissions laboratory at Victoria Park; a technical officer and lecturer at UTS, and a visiting scientist at the National Bureau of Standards in Washington DC. He achieved the rank of Senior Principal Research Scientist at the CSIRO Division of Applied Physics followed by 12 years conducting his own scientific instrumentation business. Tony is the author of several books in the fields of materials science, physics, and engineering.

Acknowledgements

The kind contributions made by my friends and colleagues, particularly Peter Davis (dec), Ron Moss (dec), Bruce Elson are greatly appreciated. Much of the information presented here was taught as part of the Sydney Technical College Automotive Engineering Certificate course whose champion was Mr Brian Le Grice. He and his colleagues, by virtue of their expert tuition, provided a unique and lifelong asset for their grateful students.

CHAPTER 1.
PRODUCTION & IDENTIFICATION

1.1 BMC Australia Product Engineering Documentation[1]

1.1.1 Sketch

In the process of vehicle design at BMC, the first engineering document is usually a sketch made by an engineer or draftsman. In the technical drawings register, they are identified with the number prefix YS, indicating "Australian Sketch".

A typical sketch is one shown in Fig. 1.1.1 for a proposed change to the routing of the handbrake cable for P76 Saloon, dated early 1974.

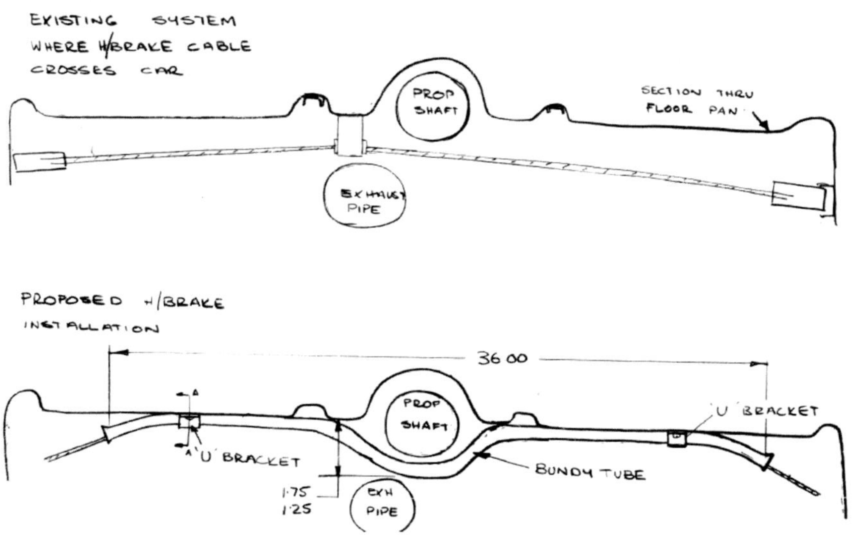

Fig. 1.1.1 *YS885, 4/74.*

If the proposed sketch is to proceed, an engineer will design the part and things move on to an Engineering Drawing.

Australian produced drawings have drawing numbers with a Y in the second character position. UK drawings and parts are either of the "BMC" format (e.g. BMK4268) or the Austin system (e.g. 22G2450). There is a set procedure for assigning drawing numbers according to the type of part (mechanical, trim, body, etc). Often as not, the Drawing Number for a single item becomes the Part Number used in production and service.

[1] Acknowledgement: Peter J. Davis

1.1.2 Part Numbers

1.1.2.1 Introduction

Part numbers used in Australia are a mixture of Austin, Nuffield (MG, Morris, Riley, and Wolseley) and Pressed Steel Fisher systems. Upon the formation of BMC, existing parts were not renumbered, new parts were identified by the current numbering system applying at the time. However, in Australia, part numbers issued before the BMC system came into operation (namely AC Australian Component) were renumbered.

Where a part is drawn on a single drawing, the Drawing Number becomes the Part Number. In many cases, a drawing may contain more than one Part Number which case an index card for that part is created as the cross reference.

Part numbers consist of two groups of characters. In the first group, the Austin part numbering system uses one or two numeric characters, followed by a single alpha character. In the BMC part numbering system, the first group comprises three alpha characters.

In the Austin system, for the first group, the first numeric character is a continuation of ranges as each block of 9999 numerics of the second group is used up. In the first group, the second numeric character denotes the division between Engine, Chassis or Body components.

In the BMC system, the first alpha character A to F (omitting G) signifies a mechanical component. Characters H to Z (omitting I, O, Q, and S) signify a body component. The second alpha character signifies the origin of the part. For Australia, the character Y is assigned.

A	Morris Commercial Vehicles, Adderley Park	Chassis
B	Morris Commercial Vehicles, Adderley Park	Body
C	Morris, Wolseley, M.G. and Riley Cars, Cowley	Chassis
D	Morris, Wolseley, M.G. and Riley Cars, Cowley	Body
E	Engines Branch, Coventry	
F	Morris Bodies Branch, Coventry	
H	M.G., Abingdon	
J	Cowley Service	Chassis
K	Cowley Service	Body
L	Fisher and Ludlow	
M	Morris Commercial Vehicles, Adderley Park	Chassis
N	Morris Commercial Vehicles, Adderley Park	Body
P	N.M.P., Ward End	Body
R	Radiators Branch, Oxford	
T	T&T Branch, Ward End	Chassis
U	S.U. Carburetters, Birmingham	
Y	BMC/Leyland Motor Corporation, Australia	
Z	Pressed Steel Company	

For both the Austin and BMC systems, the third alpha character signifies model range:

A	7-8 hp	G	10 hp
B	16 HP	H	12 hp
C	20 HP	J	Tractors
D	25 hp	K	Trucks
E	18 hp	L	Under 7 hp
F	14 hp	O	Outside contracts

In later coding, the third alpha character "A", "B", "D", "G", "H", "L" signifies engine size.

A 0800 to 0999 cc
B 2000 to 2999 cc
D 3000 cc 3999 cc
G 1000 to 1399 cc
H 1400 to 1999 cc
L 0000 to 0799 cc

1.1.2.2 Body components

In the Austin system, the four numeric characters of the second group are divided into separate thousands, e.g. 1000, 2000, 3000, ... 9000 series.

Within each of these series, the breakdown is as follows:

1	Body Shell Track
2	Paint Track
3	Finishing Track
4	Chassis Track
5	Mounting Track
6	Car Finishing Track
7	
8	
9	
10-99	Major Assemblies to Body Shell Track
100-109	Major Assemblies to Paint Track
110-259	Major Assemblies to Finishing Track
260-279	Major Assemblies to Chassis Track
280-339	Major Assemblies to Mounting Track
340-349	Major Assemblies to Car Finishing Track
450-900	Detail Part Numbers
901-999	Track Only (continuation of 1 to 9)

Drawings are only issued in the range 450 – 900 in each of the series, the remaining numbers are assigned for scheduling and costing purposes in conjunction with the Planning Department.

In the BMC system, the first alpha character, H to Z (omitting I, O, Q, and S) signifies a body component. The second alpha character signifies the origin of the part as described previously. For Australia, the character Y is assigned.

The third alpha character "A", "B", "D", "G", "H", "L" signifies engine size as described previously, but characters "A", "B", "C", "D", "E" later signified model range.

A No model significance
B No model significance
C Leyland P76
D Leyland Marina
E Leyland Mini, Moke

The numeric part is even for right handed parts, and odd for left handed parts consecutively upwards from the even numbered part.

1.1.2.3 Mechanical Components

In the BMC system, the first alpha character A to F (omitting G) signifies a mechanical component. The second alpha character signifies the origin of the part. For Australia, the character Y is assigned.

The third alpha character “A”, “B”, “D”, “G”, “H”, “L” signifies engine size as detailed previously.

In the second group, the four numeric characters 0001 to 9999 are divided:

0001 – 1999	Engine
2000 – 2999	Engine accessories (Petrol tank, radiator, exhaust system, etc)
3000 – 3999	Gearbox
4000 – 4999	Front suspension
5000 – 5999	Chassis, Handbrake, Controls
6000 – 6999	Steering
7000 – 7999	Rear axle
8000 – 8250	Road wheels
8251 – 8999	Continuation of 5000 – 5999
9000 – 9999	Electrical components (not included in the engine assembly)

The numeric is even for right handed parts, and odd for left handed parts consecutive upwards from the even numbered part.

1.1.2.4 Pressed Steel Fisher

BMC (UK) allocated the second alpha character “Z” to signify Pressed Steel Company (PSC) and later Pressed Steel Fisher (PSF).

PSC/PSF numbering system which was not converted to the BMC system. The PSC/PSF system was five numeric characters/four numeric characters. The first numeric character signified:

1	Pressing
2	Sub Assembly
3	Assembly

The second, third, fourth and fifth character positions numeric characters signified model. (e.g. 1450 was MGB).

The first, second, third and fourth numeric characters had the following significance:

0001	roof panel
0085	trunk lid assembly
0094	Rear wheel arch assembly (right)
0095	Rear wheel arch assembly (left)

These PSC/PSF part numbers were then allocated BMC part numbers using the alpha character “Z” as the second digit.

The three alpha characters “H”, “Z”. “A” were allocated to the MGB.

Some examples:

31450/0085	HZA0419 trunk lid assembly
31450/0094	HZA0422 Rear wheel arch assembly (Right)
31450/0095	HZA0423 Rear wheel arch assembly (Left)

The four numeric characters had no significance.

These PSC/PSF part numbers were mainly used within BMC and they referred to body-in-white assembly components.

1.1.2.5 Standard Parts

For standard parts, the part numbers are issued by the Standards Department. In most cases, a standard part can be identified by the third alpha character as being outside the range A to L.

1.1.3 Registers

Part Numbers are booked out in registers. Often the registers are hand-written and a small number of them survive.

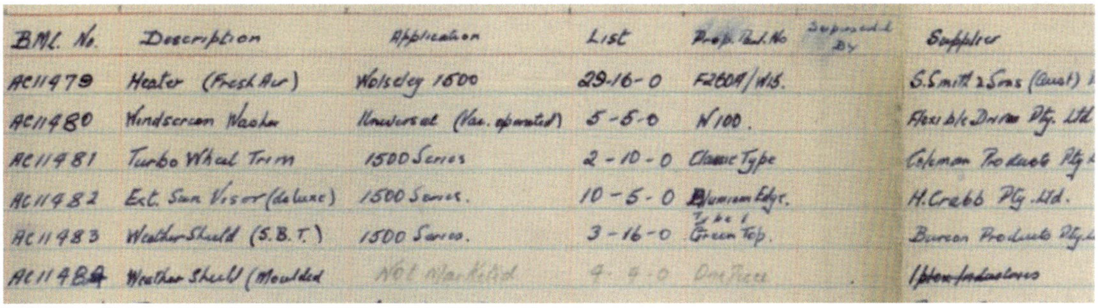

B.M.C. No.	Description	Application	List	Prop. Part No	Superseded By	Supplier
AC11479	Heater (Fresh Air)	Wolseley 1500	29-16-0	F260A/WS.		S.Smith & Sons (Aust) [illegible]
AC11480	Windscreen Washer	Universal (Vac. operated)	5-5-0	W100.		Flexible Drive Pty. Ltd
AC11481	Turbo Wheel Trim	1500 Series	2-10-0	Classic Type		Coleman Products Pty [illegible]
AC11482	Ext. Sun Visor (deluxe)	1500 Series.	10-5-0	Aluminium Edge.		H.Crebb Pty. Ltd.
AC11483	Weather Shield (S.B.T.)	1500 Series.	3-16-0	Type 1 Green Top.		Burson Products Pty [illegible]
AC11484	Weather Shield (Moulded	Not Marketed	4-4-0	One Piece		[illegible]

Fig. 1.1.2 *Entries from the Australian Component (AC) Register.*

The primary reference source for all parts scheduled for production are the Engineering Drawings. Drawings include dimensions, tolerances, materials, and finish (and as will be discussed below, revisions).

1.1.4 Product Engineering Documents

A range of documentation is prepared to allow vehicles to be manufactured. These manufacturing documents include the Schedule of Parts (similar to what is usually called a Bill of Materials), Engineering Release Notes, Stop Orders, Drawing Office Instructions, Production Supplies Release, and of course, the drawings. This documentation was the responsibility of the Product Engineering Department, and their controlling documents were:

• Vehicle Specification	• Engineering Release Note (ERN)Engineering
• Model Identification	• Change Note (ECN)
• Knocked Down Allocation Schedule (KDAS)	• Engineering Change Summary (ECS)
• Knocked Down (Allocation) Note (KDM)	• Styling Australian Sheets (SAS)
• Schedule of Parts – Parts List	• Concession Request (CR)
• Drawings	• Drawing Office Instruction (DOI)
• Part Numbering	• Production Supplies Release (PSR)

Specific vehicle build documents have been destroyed, but samples of the above documents survive. A typical Engineering Change Recommendation (ECR) is shown in Fig. 1.1.3.

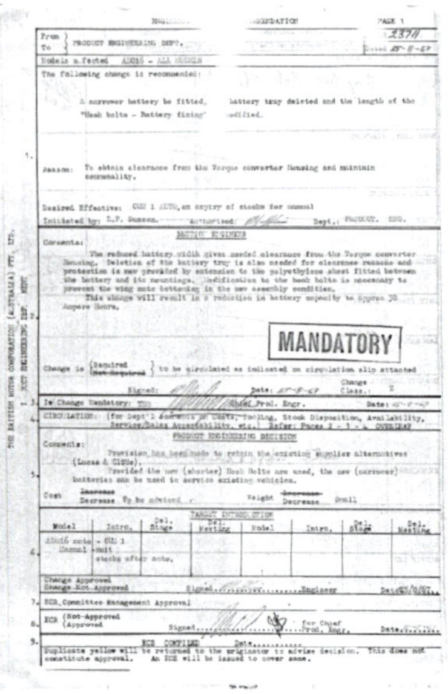

ENGI[illegible] [illegible]ENDATION PAGE 1

From / To PRODUCT ENGINEERING DEPT.	23711
Models affected AD016 - ALL MODELS	

The following change is recommended:

A narrower battery be fitted, battery tray deleted and the length of the "Hook bolts - Battery fixing" modified.

1\. Reason: To obtain clearance from the Torque converter Housing and maintain commonality.

Desired Effective: CUM 1 AUTO, on expiry of stocks for manual

Initiated by: L.F. Duncan. Authorised: Dept.: PRODUCT. ENG.

[illegible] ENGINEERS

Comments:

The reduced battery width gives needed clearance from the Torque converter Housing. Deletion of the battery tray is also needed for clearance reasons and protection is now provided by extension to the polyethylene sheet fitted between the battery and its mountings. Modification to the hook bolts is necessary to prevent the wing nuts bottoming in the new assembly condition.

This change will result in a reduction in battery capacity to approx. 35 Ampere Hours.

2\. MANDATORY

Change is Required / ~~Not Required~~ to be circulated as indicated on circulation slip attached

Signed: Date: Change Class.: [illegible]

3\. Is Change Mandatory: YES Chief Prod. Engr. Date:

4\. CIRCULATION: (for Dept'l comments on Costs, Tooling, Stock Disposition, Availability, Service/Sales Acceptability, etc.) Refer: Pages 2 - 3 - 4 OVERLEAF

PRODUCT ENGINEERING DECISION

Comments:

Provision has been made to retain the existing supplier alternatives (Lucas & Clyde).

5\. Provided the new (shorter) Hook Bolts are used, the new (narrower) batteries can be used to service existing vehicles.

Cost ~~Increase~~ Decrease To be advised. Weight ~~Increase~~ Decrease Small

TARGET INTRODUCTION

Model	Intro.	Del. Stage	Del. Meeting	Model	Intro.	Del. Stage	Del. Meeting
AD016 auto - Manual	- CUM 1 - suit stocks after auto.						

6\. Change Approved / Change Not Approved Signed........................Engineer Date

7\. ECR Committee Management Approval

8\. ECR (Not-Approved / (Approved Signed........................for Chief Prod. Engr. Date..........

9\. ECR COMPILED Date............

Duplicate yellow will be returned to the originator to advise decision. This does not constitute approval. An ECR will be issued to cover same.

THE BRITISH MOTOR CORPORATION (AUSTRALIA) PTY. LTD. 1 - [illegible] ENGINEERING DEP. MENT

Fig.1.1.3 *Engineering Change Recommendation (ECR).*

The Company Standards Engineer[2] coordinated and specified many of the engineering processes and produced a Standards Book which was distributed to some 28 different departments including Supply, Parts and Accessories, Planning, Service to name just a few. The Standards Book ensured that individual departments worked to a set of clearly identifiable specifications and covers a range of topics from vehicle identification to standard fasteners.

Fig. 1.1.4 *BMC Standards book.*

There is a set procedure for introducing engineering changes which fed back to the design process via drawing revisions. The entertaining picture shown in Fig. 1.1.5 was drawn up by the Standards Engineer to illustrate the flow of documentation.

[2] Jim Bigelow

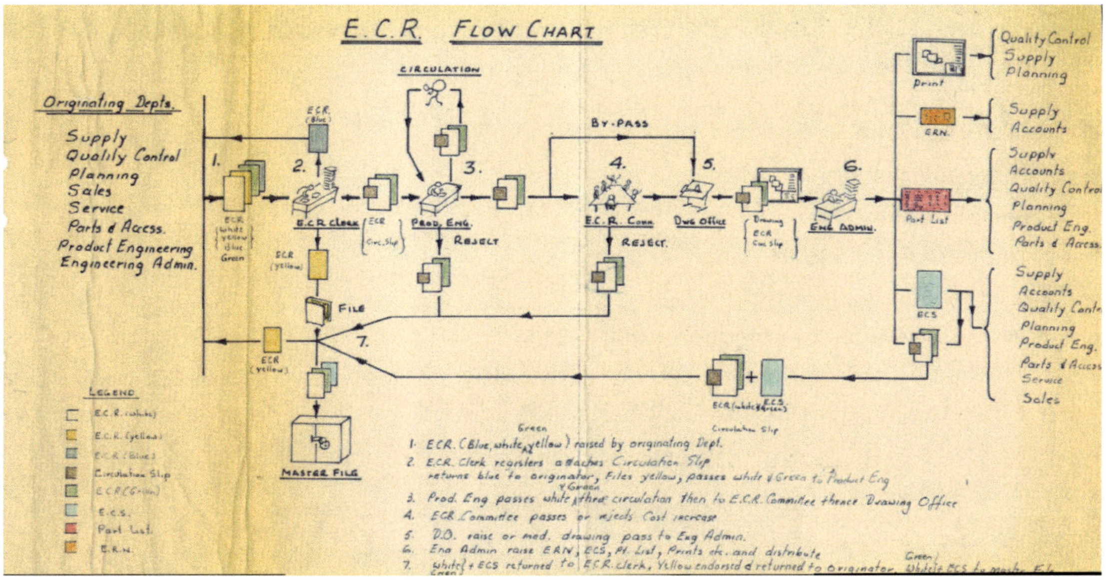

Fig. 1.1.5 *ECR flow chart.*

It will be immediately noticed that the only input to the Parts List (Schedule of Parts) are the Engineering Drawings via the Product Engineering office. Fortunately for modern day historians, a complete set of Product Engineering drawings have been preserved. A typical drawing is shown below:

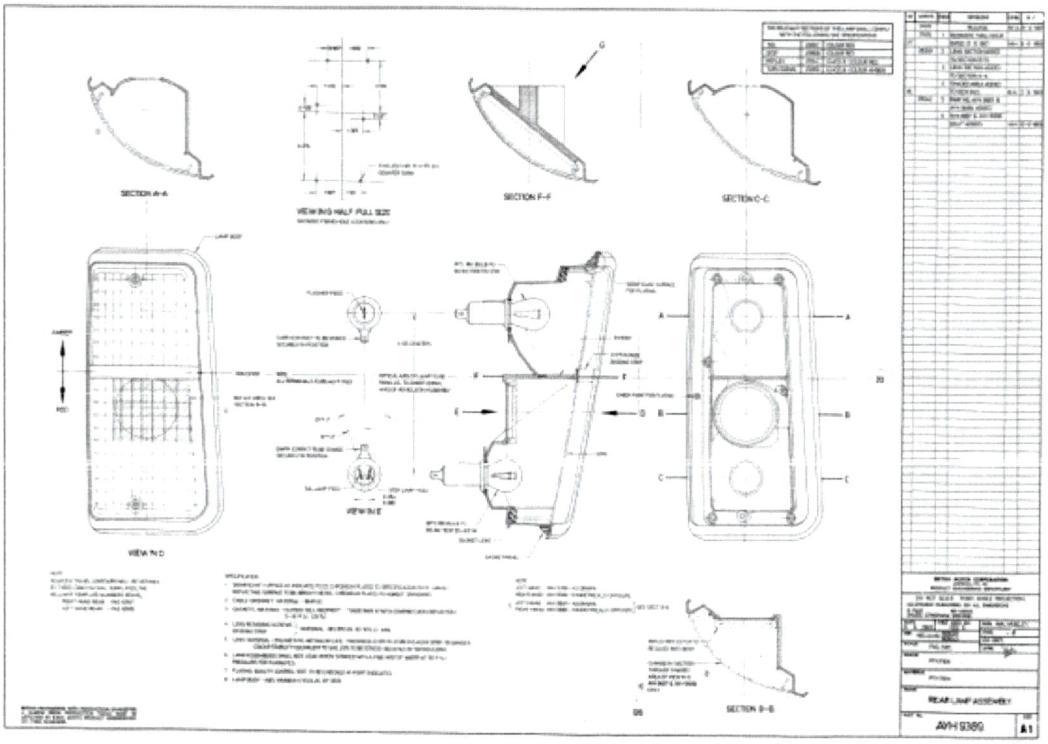

Fig. 1.1.6 *Engineering Drawing AYH9389.*

Actual manufacture proceeded via the use of planning documents prepared by Planning Engineering whose responsibilities were:

- Factory Layout
- Tooling Required
- Build Sequence
- Source of Components
- Routing of Components

Then, there is a multitude of other documents prepared later that allow the vehicles to be serviced, such as Workshop Manuals, Service Parts Lists with associated Parts Amendment Advice, Technical Bulletins, Product Problem Advice, Service Liaison Summaries, Product Fault Summaries etc.

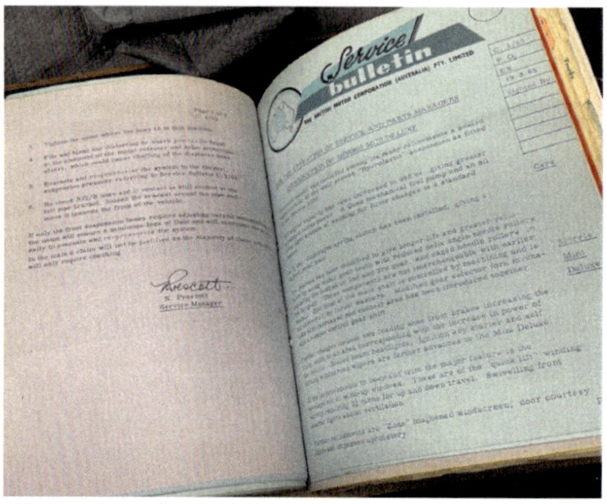

Fig. 1.1.7 *Technical Bulletins.*

The Service Parts Lists/Parts Catalogues are not catalogues for what was fitted in production (although they are often used by us now for reference purposes). Service Parts Lists were published by the Parts and Accessories Division (P&A). P&A did, however, have a copy of a subset of the Product Engineering drawings and the P&A drawings set are still in existence. In many cases, P&A arranged manufacture of the part with an outside supplier different to that which supplied production line items. The Service Parts Lists show "replacement parts".

1.2 Vehicle Identification

1.2.1 Introduction

With the increasing values of classic cars, the issue of vehicle identification has now become more important than ever. The value of a particular vehicle can be influenced by thousands of dollars depending on its degree of originality. A knowledge of what was actually fitted to a vehicle at the factory is an essential ingredient in establishing market value. In this Section, the Mini and Moke range of vehicles is used as an example of the way in which Product Engineering identified production vehicles.

For some years now, there have been published listings of identification details for the Mini and Moke range of vehicles based upon observations made of Identification Plates, Compliance Plates, passport to service details and sales dockets. While an admirable undertaking, this has often resulted in misunderstandings of just how the company (BMC (Australia), BLMC (Australia) and then Leyland Australia actually identified their products and as a consequence, led to inaccurate interpretation of the codes used.

The bedrock of vehicle identification is the documentation produced by the Product Engineering Department which, amongst other things, formulated and/or implemented identification coding for both bodies and power units. From these specifications, individual cars were coded and then stamped before being supplied to the selling dealers. Service Parts Lists are produced after-the-fact and rely on both Parts and Accessories part numbers and production part numbers.

1.2.2 Sheet Metal and Body Division Stampings

Fig. 1.2.1 *SMBD Body Stamping showina model and Cum.*

The first body stamping applied to a vehicle is done in the Sheet Metal and Body Division (SMBD) where individual bodies, when built up into what is called a body-in-white, are stamped with what is called the "Cum" (pronounced "cume"). The Cum was a cumulative, or running total of bodies produced by the SMBD so that production tallies could be recorded at the end of each shift by simply subtracting the new Cum from the old Cum. This number begins at 1 for each model. It was not supposed to be generally visible on the car when it was completed, but now leads to considerable confusion. For a Morris Mini, the Cum can often be found on the radiator shroud panel along with other stampings which indicate the model code. These model code stampings identify the marque and the body style. For example, a Morris Mini Cooper S is stamped M2/06 and then followed by the Cum. In this example, the vehicle is a Morris 2 door, model YDO6. Variations on this code can be seen in practice. In the picture here, the M was not stamped.

1.2.3 Terminology

The Cum is not a "Body Number" and it is not a "Chassis Number". Although these terms may be used by some registration authorities, they mean different things. Prior to about 1960, most passenger vehicles were built with a separate chassis upon which was affixed the body. In later years, unitary construction largely combined the functions of the chassis into the body structure. In modern times, "Chassis Number" is more correctly applied to commercial truck vehicles which do have a chassis, and upon which the identification number is stamped or an identification plate attached.

In the BMC range, while the front wheel drive models may indeed have subframes which resemble a chassis in some respects, no identifying number for the vehicle is stamped in them.

The Cum is probably the closest thing to what we might call a Body Number since it applies to the body-in-white and not a complete vehicle.

"Car No" is the term used by BMC/Leyland to individually identify a vehicle within a model range and is usually stamped in the dash panel or other metalwork visible from the engine bay. As a carry-over from historical use, the term Chassis No. may still be found (e.g. on a Morris Mini Cooper, the UK-assigned "Chassis No" appears in addition to the Australian Car No.).

Similarly, the commonly used term "Engine Number" is more correctly a "Power Unit Number". The coding used for power units in BMC/Leyland vehicles contains information not only about the engine, but also the transmission, final drive, and in some cases, even the inboard universal/constant velocity joints. However, Engine Number is the commonly used term, even on the printing of Identification Plates fitted by the factory.

1.2.4 Vehicle Identification

In general, there are two types of plates riveted to the bodies. The first is the black printed silver aluminium Identification Plate, and the second, an aluminium Compliance Plate. Commercial vehicles like Mini Van and Moke had other plates also fitted, but the ones that are stamped with identification coding are of particular interest to us here.

Identification Plates and Compliance Plates generally contained codes for the Vehicle Model (Type Code), Engine (Power Unit), and Body Colour. Additional information appears for Passenger Vehicle Derivatives.

1.2.4.1 Vehicle Identification Plates (Passenger Vehicles)

For the Mini range, from 1961 to about March 1969, vehicles were fitted with an Identification Plate.

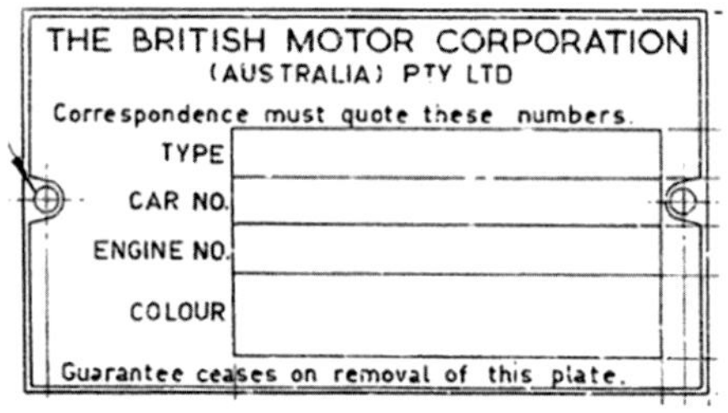

Fig. 1.2.2 *Identification Plate layout (Passenger Vehicles).*

The Identification Plate carried the following information:

Identifier	Interpretation
Type	BMC-A Type Code which follows the UK Type Code specification.
Car No.	Car serial number. To avoid giving the impression that the customer was not purchasing a very early, and potentially under-developed car, body serial numbers nearly always started at some high number, like 501, or 1001. The Car No. should not be referred to as a Body Number or a Chassis Number to avoid confusion with the SMBD CUM stampings. Further, although Minis and Mokes are fitted with subframes, they are not considered to be constructed using a chassis and the term Chassis No. is therefore somewhat misleading.
Engine No.	This field embodies a Power Unit Code followed by a Serial Number. Note, the Power unit Code should not really be referred to as the Engine Number (even if it is on the factory Identification Plate), but it is common to do so. The Power Unit code embodies information about the engine and the transmission together as a unit. UK power unit (CBU) Serial Numbers begin at 101, whereas CKD power units begin at 1001.
Colour	Paint Type: 1 – Enamel, 2 – Lacquer Colour Name: e.g. English Grey Paint Manufacturer: 1 – B.A.L.M, 2 – Lusteroid, 3 – Berger, 4 – Brolite, 5 – Taubman, 6 – Spartan

Table 1.2.1 *Identification Plate Details for Passenger Vehicles.*

From about April to December 1969, no Identification Plates were fitted to Saloon models so as to not cause confusion in the showroom, but they continued for commercial vehicles up to December 1969, whereafter the new Australian Compliance Plates were fitted. The reason for this is that the Company did not want their volume selling cars to be in a position where vehicles which could be taken to be old models were appearing in the showroom alongside vehicles which could be considered newer. A customer might ask why one car has a black plate while another has a silver plate. Plus, the new Compliance Plate had the month and year of manufacture stamped on it which made the situation worse.

1.2.4.2 Vehicle Identification Plates (Passenger Vehicle Derivatives)

The Identification Plate for Passenger Vehicle Derivatives is similar to that above but contains extra entries for GVW (Gross Vehicle Weight) and GCW (Gross Combined Weight). A typical example is:

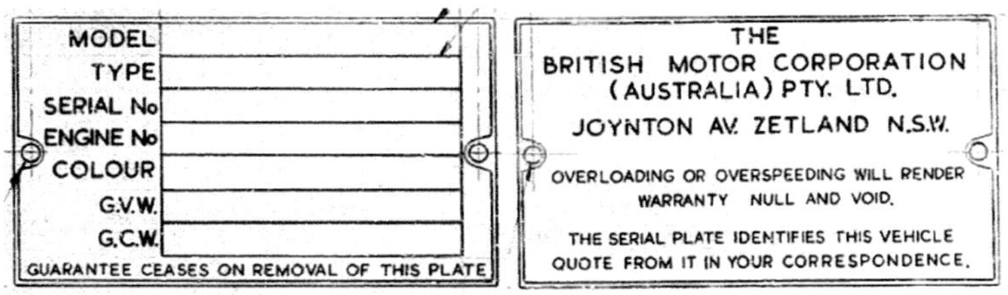

Fig. 1.2.3 *Identification Plate layout (Passenger Vehicle Derivatives).*

1.2.5 Compliance Plates

Compliance Plates were made from aluminium and fixed to the body with rivets in the same manner as for the previous Identification Plates, but in a different location. They were fitted to both Passenger Vehicles and Passenger Vehicle Derivatives.

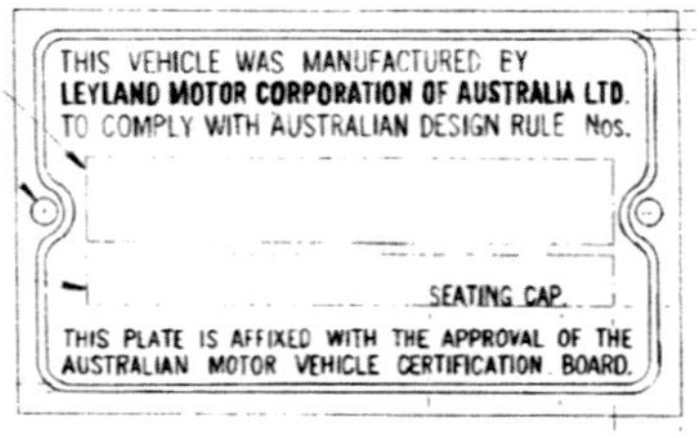

Fig. 1.2.4 *Typical Compliance Plate layout used after about 1971.*

The design of Compliance Plates changed over the years from a large square plate to a smaller rectangular plate, but essentially held the same information as above.

Apart from the Identification and Compliance Plates themselves, several coding schemes were used to identify the vehicle, both model type and power unit. These coding schemes also changed over the years and were sometimes not well understood even by those whose job it was to assign a code to a particular vehicle. The purpose of having a code in the first place is not to confuse, or keep information secret, but merely to avoid unnecessary stamping operations. For example, no manufacturer is going to stamp an engine block with "SINGLE SU, MANUAL, 4 SPEED, ALL SYNCHRO, STANDARD RATIO, DIRECT CONTROL, HIGH COMPRESSION, ELECTRICAL FUEL PUMP, PCV VALVE, EEC." when a code like 1204 would do.

The information provided is summarised in the table below.

Identifier	Meaning
Manufacturer Statement	States Vehicle Manufacturer: British Leyland Motor Corporation of Australia Ltd (January 1970 to March 1973), Leyland Motor Corporation of Australia Ltd (March 1973 to Cessation of Australian production)
Manufacturer's Designation	Model name – often inconsistently stamped, e.g. Leyland Mini, or just Mini.
Australian Design Rules	Lists applicable Australian Design Rules that applied at the time the vehicle was produced.
Vehicle Identification Code	BMC-A Type Code, Leyland Australia 10 Digit Code, or Leyland UK 8 Digit Code.
Chassis/Serial No.	BMC-A Type Codes: Continuing on from previous Vehicle Identification Plates which commenced at 501. Leyland Marque Vehicles – new models commencing at 1001. Leyland Marque Vehicles 1978 – commencing at 100001.
Date of Manufacture	Month/Year. The date at which assembly of the vehicle reaches the stage of attachment of the Compliance Plate. Generally regarded as the date at which the vehicle is considered ready for sale.
G.V.W. G.V.M.	Not present for Saloon Compliance Plates. For Passenger Vehicle Derivatives, GVW stated on a sticker in the engine bay or stamped in the Compliance Plate for later models.
Seating Capacity	Defined as the number of seated adult persons which the vehicle is designed to carry and for which seat positions are provided.
Approval statement	Australian Motor Vehicle Certification Board

Table 1.2 2 *Compliance Plate details.*

For completed vehicles, there were several different types of "Type Codes". The original BMC-A code was superseded by an Australian 10 digit code, which was then superseded by a UK 8 digit code.

1.2.6 Date of Manufacture

The Date of Manufacture is an important date in the life of a motor vehicle, and also has particular significance to registration authorities. The issue reached a level of officialdom with the introduction of the Australian Design Rules which forced manufacturers to include certain safety features, the number of which increased each year. The "Compliance Plate" listed those rules for which a vehicle was designed to meet. The Date of Manufacture was embossed on the Compliance Plate as a Month/Year combination. Compliance Plates were introduced in Jan 1970.

According to the official BMC Standards, Clause 2.5, the Date of Manufacture is defined as

> *"the date at which assembly of the vehicle reaches the stage of attachment of the Compliance Plate".*

A rather circular statement. For the time of manufacture at Victoria Park, the Compliance Plate was fitted to the vehicle towards the end of the assembly line. According to one authoritative source, the date stamping on the Compliance Plate is that which occurs when the vehicle was released from the holding yard to the selling dealer - but this is not in accordance with the Compliance Plate regulations (as they were in early 1970) even if this did indeed occur.

In 1981, when the Company was selling Rover SD1 and Jaguar products, the Built Date is defined in company documents as

> *"The Calendar month and year in which the body shell and power train sub-assemblies are conjoined and the vehicle is driven or moved from the production line."*

This date, the "Built Date" is stamped into a "Built Date" plate on the right hand side on the inner valance panel. This plate is different to the Compliance Plate.

1.2.7 Plate Attachment

It is not often appreciated that the rivets that hold the Identification Plate on to the body shell are not 1/8 aluminium rivets commonly stocked at a hardware store, but are domed head monel (nickel/copper) rivets which have a smaller head diameter than an aluminium rivet. The DMP part number prefix of the genuine part means domed-head monel pop rivet. It's a small detail, but one immediately obvious to someone who looks for accuracy when a car is judged on its originality. Of course if a plate is removed, the guarantee on the vehicle automatically ceases. Indeed, removal of an Australian Compliance Plate from a vehicle may be an offence.

Pictured below are details of construction of the Identification Plate introduced for the BMC range in 1960 and used for the next 10 years. Given the harsh environment, it a credit to their manufacturer that they have generally survived and the printed matter still readable.

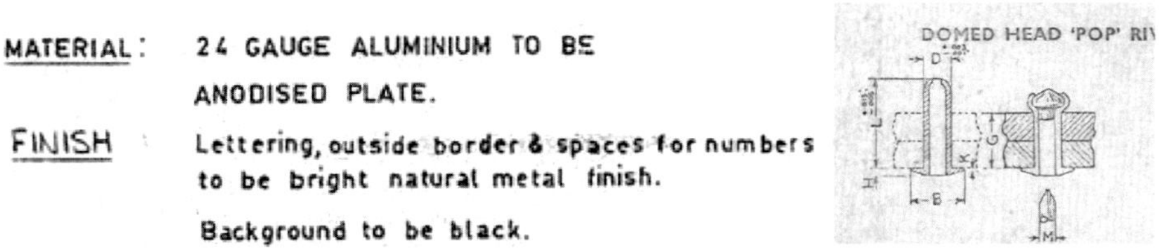

Fig. 1.2.5 *Identification Plate and pop rivet detail.*

1.2.8 The BMC-A Type Code

This Type Code follows on from the UK established practice but evolved to suit Australian manufacturing details.

Identifier	Symbol	Interpretation
Origin	Y	Australian manufacture
Marque	M K JB	Morris Cooper BMC Commercial
Engine Capacity	A G	A Series 848, 997, 998 cc G Series 1098cc, 1275cc
Body Type	2S 2V B	Saloon 2 Door Van 2 Door Buckboard
Series of Model	Numeric	Commencing at 1

Table 1.2.3 *BMC-A Type Code.*

It was used for over a decade on the Mini and Moke range from the Morris 850 to the Morris Mini Clubman.

1.2.9 The Leyland Australia 10 Digit Code

The Leyland Australia 10 Digit Prefix Code was an internally developed code which was first proposed in Production Engineering as far back as 1969, but not adopted until January 1973. This new code became necessary when it was obvious that the BMC type coding could not adequately capture the desired information about the vehicles. The new code has the following structure:

Identifier	Symbol	Interpretation
Model Code (Marque)	021	Morris Mini 1100 (Standard) (Morris Mini Clubman Saloon), Leyland Mini
	022	Morris Mini Clubman Deluxe, Leyland Mini S
	023	Morris Mini Clubman GT
	018	BMC Moke
	030	Californian Moke
Spare position	X	Later used as ADR Compliance after Jan 1973
No. Doors	0	No Doors – Buckboard (Moke)
	2	Saloon, Coupe, Van 2 Door
Body Type	B	Buckboard
	S	Saloon
	V	Van
Trim Quality Level	1	1st level of luxury
	2	2nd level of luxury
	3	3rd level of luxury
	4	4th level of luxury
Transmission	M	Floor Shift Manual
Engine Capacity	09	A Series 4 cylinder 0900-0999cc
	10	A Series 4 cylinder 1000-1099cc
	12	A Series 4 cylinder 1200-1299cc

Table 1.2.4 *Leyland Australia 10 Digit Code.*

The code embodied a spare position (later to be used for ADR listings); mnemonic abbreviations which were supposedly easy to remember without reference to spec sheets; specification segments alternatively using numerical and alphabetical characters; the main model identification being the first 3 characters; and the engine and transmission features being the last three characters. It was a very good system when correctly used.

YDO21 and YDO22 Body Stamping Problem

Problems arose when there was a change in Quality Control personnel who assigned the code to individual vehicles and interpretation was done differently. No more is this in evidence than in the Mini range YDO21 and YDO22 (Morris Mini Clubman and Leyland Mini).

As is well known, YDO21 is the standard model Morris Mini Clubman (Standard) and Morris Mini Clubman Van, while YDO22 applies to the more luxurious Morris Mini Clubman Deluxe, Leyland Mini S, Leyland Mini SS, Leyland Mini Sunshine, Leyland Mini LS.

Just focussing on the saloon models, YDO21 should be a base model Mini, originally fitted with a centre gear change (like the Morris 850), centre speedometer, rubber floor mats, etc. The there is only one "luxury" level for this car, and that is a rather basic level of trim when compared to YDO22. Accordingly, the Type Code for YDO21 with a 1098cc power unit should be 021X2S1M10 (where X has been written as a place holder for the ADR listing code which doesn't concern us just now).

However, observations show that the cars were stamped 0212S2M10 which implies that there was even a lower level of trim available when there wasn't.

A similar mistake occurs with YDO22. These vehicles should be stamped 022X2S1M10. Why "1" you say? Because originally (before the Leyland Mini SS, Sunshine and LS models) there was only one level of trim for YDO22 and even though it was more luxurious than that fitted to YDO21, for the YDO22 model, it was the lowest (and only) level of trim available. But, someone got things mixed up and stamped these cars 022X2S2M10 which often goes un-noticed because one thinks that the "2" signifies the more luxurious Leyland Mini S compared to the basic standard Leyland Mini. This is incorrect. YDO21 and YDO22, despite their obvious similarities, are different models. The trim level applies *to that model only*, and so a Leyland Mini S, should be stamped 022X2S1M10.

In later years, a Leyland Mini Sunshine would be stamped 2S2, a Leyland Mini SS would be stamped 2S3, and a Leyland Mini LS, 2S4 which follows the correct procedure. Confusingly, at Enfield, someone caught on to these mistakes and Leyland Mini S started being stamped 2S1 which by coincidence, happened about the same time as the introduction of the UK CBU 998 power units, which by a further coincidence corresponded soon after the move of manufacturing to Enfield and a change of Quality Control manager.

1.2.10 The Leyland UK 8 Digit Code

From March 1978, Leyland Australia used the UK-assigned 8 Digit Prefix Code. It should be noted at the outset that Australia was not part of the European Economic Community and therefore was not required to have vehicles stamped with a VIN number. A VIN number is a 17 digit code in total. The code used by Leyland Australia does not include all the VIN fields and it is not precisely a VIN number - although it is referred to as such in company literature. The 8 Digit Code – for the fields in encompasses – is, however, similar to the VIN standard as used in UK.

Identifier	Symbol	Interpretation
Model Code (Marque)	XN AK	Leyland Mini Leyland Moke
Model Class	F H P	High Line Mid Line Low Line
Body Style	A F P	Saloon 2 Door Van 2 Door Buckboard (Moke)
Engine Capacity	B D	A Series 998cc A Series 1275cc
Transmission/Drive	1 2	Manual, 4 Speed, RHD Manual, 4 Speed, LHD
Year of Model Introduction	8 9	1978 1979
Country of Origin	Y	Australia
Car Serial No.	Numeric	Commencing at 100001

Table 11.2.5 *Leyland UK 8 Digit code.*

1.3 Power Unit Identification

1.3.1 Introduction

A variety of power units with differing engine capacities, transmission specifications, final drive and other mechanical items are fitted to the BMC range. The coding used for power unit identification changed over the years as it became necessary to distinguish the ever-increasing complexity of vehicle specifications.

In general, for the Mini and Moke range, there are four basic types of Engine Prefix Codes:

- UK Code (e.g. 8AM|U|H)
- BMC Australia (e.g. 8Y|U|H)
- Australian 4 Digit (e.g. 1206)
- British Leyland UK (e.g.99H905AJH)

In UK, power units were identified with stampings made on a plate which was then riveted to a pad on the engine block casting. This method of attachment was not suitable for Australian registration requirements, and so the UK plate (where supplied) was removed, and the identification code stamped directly into the pad.

Fig. 1.3.1 *Example of a UK power unit Identification Plate.*

Fig. 1.3.2 *Example of an Australian power unit identification stamping.*

1.3.2 BMC-A Power Unit Coding

The BMC-A Engine Code Prefix used the UK system with a series of codes to indicate the engine capacity, origin, transmission, compression. The Prefix was followed by the Serial Number which uniquely identified a particular power unit.

Identifier	Symbol	Interpretation
Capacity	8	848cc
	9	997, 998, 1275cc
	10	1098cc
	12	1275cc
Origin	AM	Austin-Morris (UK)
	Y	Australia
Power Unit Variant	Alpha	Commencing at A
Transmission	A	Automatic
	U	Centre Gear Change
	Sa	Close Ratio Gearbox Remote Control
	Ta	Standard Ratio Gearbox Remote Control Three Speed Synchro
	Te	Standard Ratio Gearbox Remote Control Four Speed Synchro
	We	Standard Ratio Gearbox Centre Gear Change Four Speed Synchro
	Xe	Close Ratio Gearbox Remote Control
Compression	H	High
	L	Low
Serial No.	Numeric	Commencing at 1001

Table 1.3.1 *BMC-A Engine Prefix Codes and Serial Number.*

Although the above table signifies the transmission code specifications, a mistake was made at the factory where engines with four speed synchromesh transmissions continued to be stamped with the code Ta and U, where they should have been stamped Te and We respectively.

Note, there are two types of "B" form gears. The B form gears have a helix angle (for the helical gears) different to the earlier ones, but some gear sets were made in Australia by Borg Warner and these are identified by a B in the engine number. These locally made gears that may not be compatible with B form gears made in UK. It is said that the locally made Borg Warner gears were only used on Morris 1100 range.

1.3.3 Australian 4 Digit Prefix Code

This simplified code was intended to reduce the number of stampings while at the same time increasing the number of possible variations. The consequence was that the codes could not be easily remembered.

The first two digits of the 4 Digit Code signified the engine capacity.

Engine capacity	Two-digit code
0900 – 0999	09
1000 – 1099	10
1200 – 1299	12

Table 1.3.2 *Engine capacity code.*

The second two digits signify the power unit variant.

Some examples are:

Engine No. Prefix	Power Unit	Design Reference
1000	Single SU, Manual, 4 Speed Synchro, Standard Ratio, Direct Control, High Compression, Mechanical Fuel Pump, PCV Valve	YDO21 YDO21V
1001	Single SU, Manual, 4 Speed Synchro, Standard Ratio, Remote Control, High Compression, Mechanical Fuel Pump, PCV Valve	YDO21 YDO22
1200	Clubman GT, Twin SU, Manual, 4 Speed Synchro, Close-Ratio, Remote Control, High Compression, Electrical Fuel pump, PCV Valve.	YDO23
1204	Single SU, Manual, 4 Speed Synchro, Standard Ratio, Direct Control, High Compression, Electrical Fuel pump, PCV Valve, EEC.	YDO30 YDO30A
1206	Clubman GT, Main Bearing Cap Change, Twin SU, Manual, 4 Speed Synchro, Close-Ratio, Remote Control, High Compression, Electrical Fuel pump, PCV Valve.	YDO23

Table 1.3.3 *Power Unit Code variants.*

1.3.4 British Leyland UK Code

When production of the Mini and Moke range was relocated to Enfield, once the stock of 1098cc engines produced at the Waterloo site had been used up, power units were then imported from UK as CBU items, and as a consequence, reverted from 1098cc to 998cc (with later addition of 1275 CBU power unit as an option for the Moke Californian and Leyland Mini LS).

A UK assigned Power Unit Prefix Code was introduced in 1970 in UK and in about 1975 in Australia on some models and combines the earlier 2nd Prefix code into a series of three numerics followed by a letter code of one to three characters.

The first prefix code signifies the engine capacity and orientation:

Code	Description
99H	998cc Transverse
10H	1098cc Transverse (not used in Australia)
12H	1275cc Transverse

Table 1.3.4 *First Prefix Code Examples.*

The second prefix code is a numerical code which itself decoded into the previously used alphabetical codes. Some examples are given in Table 1.3.5.

1st Prefix	2nd Prefix	Coding	Model	Layout
99H	834	Yd Re Sk	YDO21	Transverse
99H	889	Yd Re Fm Sk	YDO21	Transverse
99H	905	Yd Re Bf Bm Sk Ze	YDO18	Transverse
12H	902	Yd Re Bf Yg Sk	YDO22	Transverse
12H	905	Yd Re Bf Bm Sk Ze	YDO18	Transverse

Table 1.3.5 *Second Prefix Code Examples.*

The third prefix code indicates additional features. Some examples are shown in Table 1.3.6.

Code	Interpretation
A	NO ADDITIONAL FEATURES
AJ	CCV CLEAN AIR PACK EVAPORATIVE LOSS CONTROL 2ND ALTERNATIVE SINGLE CARBURETTER

Table 1.3.6 *Third Prefix Code Examples.*

A final character code indicates an engine condition:

Code	Interpretation
H	HIGH COMPRESSION PETROL
L	LOW COMPRESSION PETROL
X	970 CC COOPER SPECIAL ONLY
Y	1275CC COOPER SPECIAL ONLY
Z	A SERIES PLUS

Table 1.3.7 *Final Prefix Code Examples.*

For example, a power unit coded 99H860AJ would decode to a 998cc power unit with

> "MECHANICAL PETROL PUMP, EXHAUST EMISSION CONTROL, ALTERNATOR NEG EARTH, STANDARD RATIO REMOTE CONTROL GEARBOX OFFSET SPHERE JOINT, ALTERNATIVE FINAL DRIVE RATIO 4.267, CCV CLEAN AIR PACK EVAPORATIVE LOSS CONTROL 2ND ALTERNATIVE CARBURETTER"

The term "offset sphere joint" means what is generally called a "pot joint" – an inboard constant velocity joint as distinct from a rubber cross or Hardy-Spicer universal joint.

1.4 Vehicle Identification Standards

1.4.1 Introduction

The Product Engineering Department in Australia was responsible for vehicle identification and the specifications for the procedures are documented in Standards published internally for company use.

Naturally, these Standards are modelled closely on those issued by UK Engineering Departments but there were some departures for local conditions. For example, individual YDO vehicle identification numbers were issued by the Australian department, as well as the 10 digit Leyland Australia Type Code.

Many of these Standards still exist and are an excellent resource for historians interested in vehicle identification at BMC and Leyland Australia.

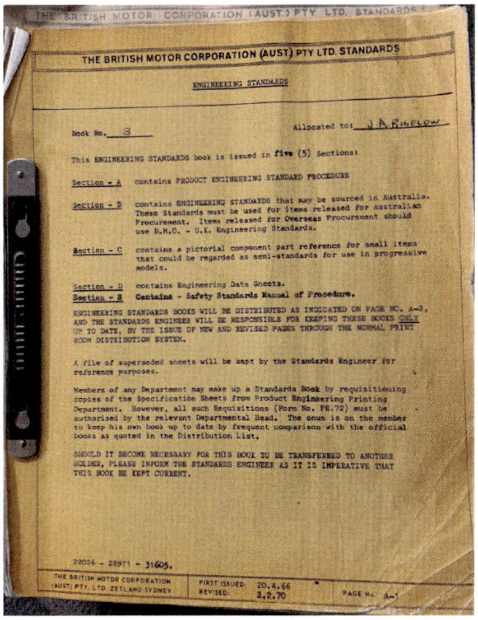

THE BRITISH MOTOR CORPORATION (AUST.) PTY LTD. STANDARDS

THE BRITISH MOTOR CORPORATION (AUST) PTY LTD. STANDARDS

ENGINEERING STANDARDS

Book No. 3 Allocated to: J.A. Bigelow

This ENGINEERING STANDARDS book is issued in **five** (5) Sections:

Section - A contains PRODUCT ENGINEERING STANDARD PROCEDURE

Section - B contains ENGINEERING STANDARDS that may be sourced in Australia. These Standards must be used for items released for Australian Procurement. Items released for Overseas Procurement should use B.M.C. - U.K. Engineering Standards.

Section - C contains a pictorial component part reference for small items that could be regarded as semi-standards for use in progressive models.

Section - D contains Engineering Data Sheets.

Section - E Contains - Safety Standards Manual of Procedure.

ENGINEERING STANDARDS BOOKS WILL BE DISTRIBUTED AS INDICATED ON PAGE NO. A-2, AND THE STANDARDS ENGINEER WILL BE RESPONSIBLE FOR KEEPING THESE BOOKS ONLY UP TO DATE, BY THE ISSUE OF NEW AND REVISED PAGES THROUGH THE NORMAL PRINT ROOM DISTRIBUTION SYSTEM.

A file of superseded sheets will be kept by the Standards Engineer for reference purposes.

Members of any Department may make up a Standards Book by requisitioning copies of the Specification Sheets from Product Engineering Printing Department. However, all such Requisitions (Form No. PE.72) must be authorised by the relevant Departmental Head. The onus is on the member to keep his own book up to date by frequent comparison with the official books as quoted in the Distribution List.

SHOULD IT BECOME NECESSARY FOR THIS BOOK TO BE TRANSFERRED TO ANOTHER HOLDER, PLEASE INFORM THE STANDARDS ENGINEER AS IT IS IMPERATIVE THAT THIS BOOK BE KEPT CURRENT.

22026 - 28971 - 31605.

THE BRITISH MOTOR CORPORATION (AUST) PTY. LTD. ZETLAND SYDNEY	FIRST ISSUED: 20.4.66 REVISED: 2.2.70	PAGE No. A-1

Fig. 1.4.1 *Front page of BMC (Australia) Engineering Standards issued 1966.*

1.4.2 Engineering Standards (Australia)

Part 1 of this Standard deals with Product Engineering standard procedures and includes Vehicle Identification as well as Parts Lists; Part Numbering System; internal Production documents such as Engineering Release Notes, Stop Orders, and Process Specifications.

1.4.3 Engineering Technical Data

Engineering Technical Data sheets are a detailed listing of specifications, both body and mechanical, for each model produced. They represent a complete description of basic specifications, additional items, special production orders, for individual models.

1.4.4 Standard Procedure Instructions

Standard Procedure Instructions give many details about Company operations from an upper management perspective and include details of major changes to Vehicle Identification schemes, such as the introduction of the 10 digit code.

1.4.5 Service Documentation

Service Liaison Summaries, Drawing Office Instructions, Service Bulletins, Technical Bulletins, Product Problem Advice, Product Fault Summaries, Service Parts Lists, and Workshop Manuals also give details of Vehicle Identification from different perspectives which help detail changes to vehicle specification as production proceeded.

POWER UNIT SERIAL NUMBERS.

SECOND PREFIX GROUP

THIRD GROUP

Fig. 1.4.2 *British Leyland Standards (Longbridge) 1968.*

1.4.6 UK Standards

In early years, Product Engineering in Australia relied heavily on established UK practice. During the 1960s and 1970s, the Australian Product Engineering department was somewhat autonomous. Somewhat ironically, towards the end of Australian production, the reliance on UK standards returned with the introduction of world-wide model type codes and engine numbering systems. UK Standards were issued by British Leyland Standards Department, Longbridge, in Birmingham UK.

1.5 Production Numbers

1.5.1 Introduction

Of particular interest to Mini and Moke enthusiasts is the question of how many of each model were built. This is a difficult question to answer, and one which engenders considerable acrimonious debate amongst enthusiasts.

1.5.2 Mini

In this section, a different approach to that taken by others is presented where actual factory figures are referred to as they have been discovered. Not all models can be treated in this way because some data has been aggregated before being tabulated in factory documents, but an overall picture can be gained.

For the Morris Mini range, there is excellent data available from the personal records made by Major John Frew. Frew was known to G.A. Lloyd and after his rather extensive and traumatic war experience, Lloyd told Service Manager Tom Poole to find Frew a job at the factory. Included in a wealth of technical documentation he amassed, Frew personally recorded the Type Code, Car No, and Engine No. for all production cars at the beginning of each year starting in 1955. He retired from BMC in the early 1970s.

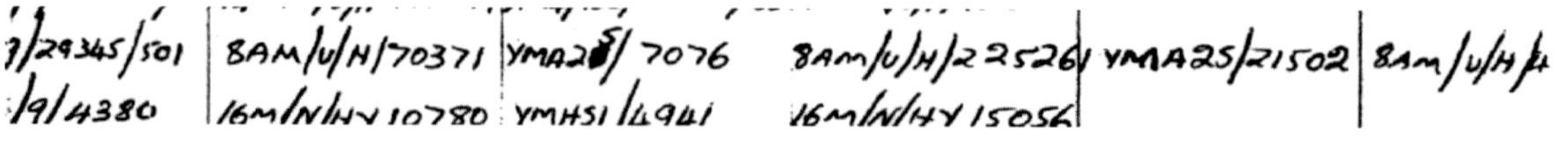

Fig. 1.5.1 *Sample of John Frew's hand-written record showing first production ADO15.*

Model	Type Code	Car No.	Engine No.
1961			
ADO15 Morris 850 (Mar)	YMA2S1	9|29345|501	8AM|U|H70371
1962			
ADO15 Morris 850	YMA2S1	7076	8AM|U|H225261
1963			
ADO15 Morris 850	YMA2S1	21502	8AM|U|H403954
1964			
No data			
1965			
YDO5 Morris Mini Deluxe (Mar)	YMA2S2	501	9YA|Ta|H1003
YDO6 Morris Cooper S (Jul)	YKG2S2	501	9F/Sa/Y36207
ADO15 Morris 850 (Nov)	YMA2S1	60039	8Y|U|H22068
ADO15 Morris 850 (to Jan 66)[1]	YMA2S1	60688	8Y|U|H23094
1966			
YDO5 Morris Mini Deluxe (Jan)	YMA2S2	13427	9YA|Ta|H14173
YDO6 Morris Cooper S (Jan)	YKG2S2	816	9F/Sa/Y37585

Note 1. ADO15 fitted with wind-up windows.

Table 1.5.1 *John Frew's production data for the Morris Mini range.*

A useful booklet produced by Wheels Magazine as part of the *"How to get more from your car"* series appeared in about 1969.

Some of the data presented are a little doubtful, but some are in agreement with other published data. For example, the total given for Morris 850 appears identical with that of John Frew. The data for Morris Cooper S MKI is identical to that provided by Norwood (see below). On the other hand, the data for Mini Van ADO15/V for just over one year's production shows 3984, while for YDO4/V, over four years, shows 3948 units.

AUSTRALIAN MINI PRODUCTION

Model	Introduced/Discontinued	No Produced
Morris 850	January 1961/January 1966	60,188
*Morris 850 (a)	February 1966/February 1969	2,248
*Morris 850 (b)	March 1969/continues	
Morris Mini Deluxe	March 1965/March 1969	37,489
Morris Mini-Matic Mk I	January 1968/August 1969	2,161
Morris Mini-Matic Mk II	September 1968/continues	
Morris Mini-K	March 1969/continues	
Mini Van (sliding windows)	February 1964/April 1965	3,984
Mini Van Mk I (wind-up windows)	May 1965/February 1969	3,948
Mini Van Mk II	March 1969/continues	
Mini Moke	February 1966/March 1969	1,592
BMC Moke	April 1969/continues	
Morris Cooper 1275S Mk I	September 1965/April 1969	4,986
Morris Cooper 1275S Mk II	May 1969/continues	
Morris Cooper 997	October 1962/March 1964	3,888
Morris Cooper 998	April 1964/August 1965	

*Morris 850 (a) and (b) for fleet orders or export; (a) wind-up windows and standard suspension, (b) wind-up windows and Hydrolastic suspension.

10

Fig. 1.5.2 *"How to get more from your car", Wheels Magazine.*

In 1993, company employee G. Norwood prepared a document which was faxed to a dealer in response to an enquiry. Norwood listed YDO6 and 2/YDO6 production data. This document is considered to be reliable and is reproduced in our "It's a Mini World" book.

Although some production data may be gleaned from service documentation and registers of actual surviving vehicles, these are too unreliable to make a definitive statement on numbers. One can never be sure that registers of this type have captured the last produced vehicle, nor if they account for exports, special production orders, fleet sales and the like. Further, Service Parts Books and Technical Documentation are rife with errors and mis-types which can easily give a false account of what was produced.

A further source of information are sales figures. Interestingly, several years of sales figures can be gleaned from company documents and a consolidation of those available give the results shown in Table 1.5.1.

Mini Saloon and Van	1969-1970	1970-1971	1971-1972	1972-1973	1973-1974	1974-1975	1975-1976
Domestic	9805	8728	7993	8397	7300	8700	10700
Export	465	627	497	575			500
Total	10270	9355	8490	8972	7300	8700	11200
Moke							
Domestic	1015	1259	762	1659	1633	2100	2200
Export	1227	1283	1368	1042	1700	1800	1800
Total	2242	2542	2130	2701	3333	3900	4000

Table 1.5.1 *Consolidation of factory sales data.*

Note, sales data do not take into account the number of vehicles in stock, both at the factory and at dealers, and so the totals likely to be somewhat less (but probably by not all that much) from production figures.

Averaged over all the years reviewed, the yearly, monthly, weekly and daily totals are shown in Table 1.5.2.

Average Production				
	pa	pm	pw	pd
Mini	9184	765	191	38
Moke	2978	248	62	12

Table 1.5.2 *Average production for Mini and Moke from Sales Data.*

As mentioned previously, these averages are thought to be on the low side on account of lack of counting of stock.

CHAPTER 2.
POWER UNIT

2.1 Introduction

2.1.1 The Mechanics of Machines

2.1.1.1 The Four-Bar Mechanism

A machine is a combination of solid bodies with constrained relative motions which is used for transmitting or transforming available energy for the purpose of doing work. The components of a machine, the mechanism, are in their simplest form, a series of bars linked together.

A bar is constrained by being paired with another bar. The pairing might take the form of a pin joint, or a sliding joint, or a line contact. Some examples of the derivatives of a four-bar kinematic chain are shown in Fig. 2.1.1.

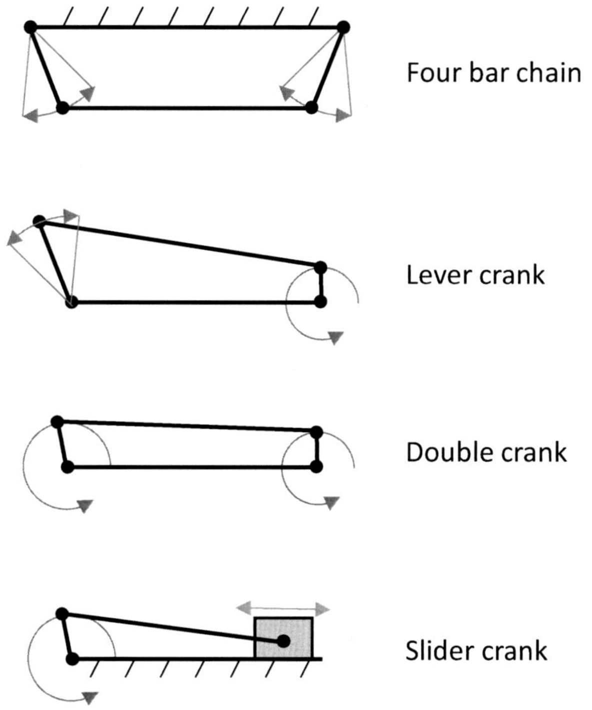

Fig. 2.1.1 *Examples of the derivatives of a four-bar chain.*

The four-bar chain is a fundamental mechanism from which all others can be derived. If any link has rotary motion, it is called a "crank". If any link has oscillatory motion, it is called a "lever". Thus, a simple four-bar chain can be one that is found on the steering mechanism of a motor vehicle. Lever-crank mechanisms can be found in the wiper motor mechanism. A double-crank mechanism can be found in the meshing gears of the transmission. A slider-crank mechanism can be found in at the engine crankshaft, conrod and piston.

An analysis of the motion of the various levers and cranks in a mechanism can be made using what are called velocity and acceleration diagrams. This is a graphical method of analysis that avoids the heavy manipulation of geometrical relationships between multiple links. Such diagrams allow the forces on the links and levers to be computed, the displacements of their ends, and many other design parameters of interest. In this book, an analysis of complicated mechanisms by this method is beyond our scope, but it is worth knowing that a formal method of the analysis of machines exists. For the mechanisms presented in this book, they are generally simple enough to take a purely analytical approach.

2.1.1.2 The Instantaneous Centre

In the analysis of reciprocating connecting rods, crankshafts and cams, the concept of an instantaneous centre if often encountered.

If a body moves in space at any particular instant, it can be shown that the motion of the body is equivalent to a rotation of the body about some fixed point in space.

Consider the instantaneous linear velocity of the points A and B which are at the ends of a rod-like body as shown in Fig. 2.1.2. Lines are drawn perpendicular to the direction of motion of each of the points. The instantaneous centre is where these lines intersect.

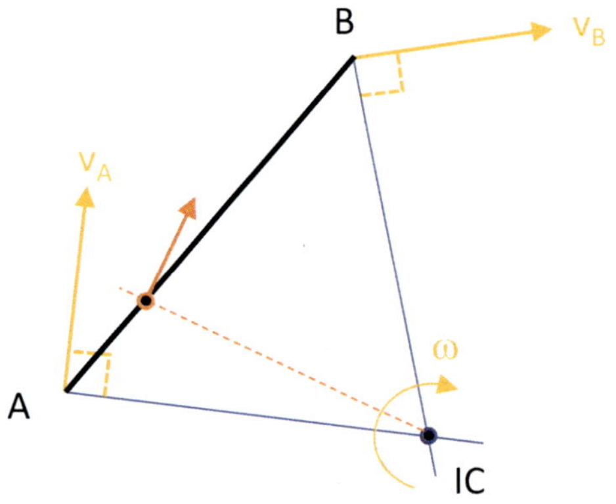

Fig. 2.1.2 *The instantaneous centre.*

The direction of the linear velocity of any point along the rod is perpendicular to the radius of that point with respect to the instantaneous centre. The magnitude of the velocity of that point is the radius of the line to the instantaneous centre multiplied by the instantaneous angular velocity ω. The instantaneous centre is used in mainly graphical methods of analysis. An example of an instantaneous radius of a cam profile is shown in Fig. 2.5.3.

2.1.2 Equations of Motion

The equations of motion for the piston in a reciprocating engine are very important for establishing the pressure, and hence forces at the gudgeon pin, the torque applied to the crankshaft, and the inertia forces associated with the reciprocating masses.

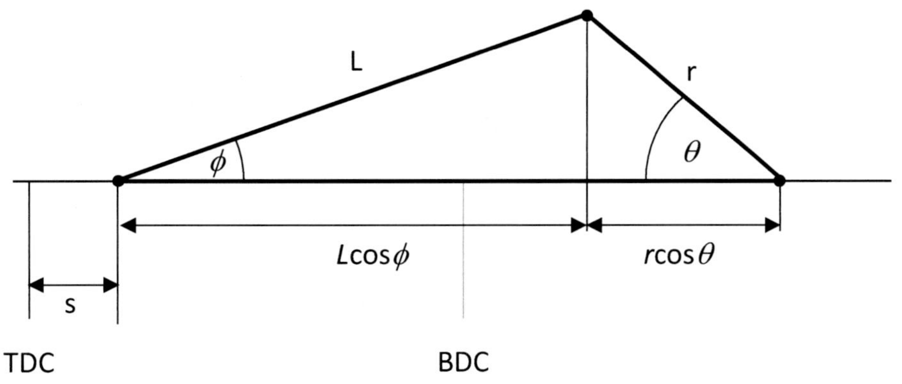

Fig. 2.1.3 *General arrangement of piston, connecting rod, and crankshaft.*

The displacement *s* from TDC position

$$s = r + L - (rcos\theta + Lcos\phi)$$

The height of the heavy triangle shown in Fig. 2.1.3 is shared between the two right-angled triangles enclosed and as such:

$$rsin\theta = Lsin\phi$$

Letting the ratio *L*/*r* = *n*, this becomes:

$$sin\phi = \frac{1}{n}sin\theta$$

Our aim is to express *s* as a function of θ alone, and so making use of the trigonometric identity:

$$sin^2\theta + cos^2\theta = 1$$

we have:

$$cos\phi = \sqrt{1 - sin^2\phi}$$
$$= \sqrt{1 - \frac{sin^2\theta}{n^2}}$$
$$\approx 1 - \frac{sin^2\theta}{2n^2}$$

And so, the displacement *s* is:

$$s = r + L - rcos\theta - L + L\frac{sin^2\theta}{2n^2}$$
$$= r(1 - cos\theta) + L\frac{sin^2\theta}{2n^2}$$

The velocity *v* is

$$v = \frac{ds}{d\theta}\frac{d\theta}{dt}$$

$$v = \omega r\left(sin\theta + \frac{sin2\theta}{2n}\right)$$

and acceleration *a* of the piston is:

$$a = \frac{dv}{d\theta}\frac{d\theta}{dt}$$
$$a = \omega^2 r\left(cos\theta + \frac{cos2\theta}{2n}\right)$$

In these equations, *n* is the ratio of the connecting rod length *L* over the crankpin length *n* and is usually about 3.5 to 4.5 for automotive applications.

Gas pressure *p* from combustion acting over the area *A* of the piston gives rise to a force on the piston, a component of which acts along the connecting rod depending upon the crankshaft angle. The component of piston force along the connecting rod acts on the crank pin to produce torque at the crankshaft. The maximum leverage occurs when the angle between the crank pin and the connecting rod is 90°. At this condition,

$$tan\theta = n$$

Table 2.1.1 shows the % distance along the stroke of the piston at the condition of maximum leverage for a range of engine capacities for the BMC A series engines.

Capacity (cc)	**850**	**998**	**1098**	**1275**	
bore	62.94	64.558	64.558	70.6	mm
stroke	68.26	76.2	83.72	81.33	mm
Conrod length	146.05	146.05	146.05	146.05	mm
n = L/r	4.28	3.83	3.49	3.59	
Crank angle	76.8	75.4	74.0	74.4	deg
s	30.1	33.1	35.9	35.0	mm
% stroke	44.2	43.5	42.8	43.0	%

Table 2.1.1 *% travel of stroke at maximum connecting rod leverage for various engine geometries A series engines.*

For a reciprocating engine, it is usually assumed that, for the sake of analysis, a third of the mass of the connecting rod is added to the mass of the reciprocating piston m_p. The inertia force arising from the reciprocating mass m_p at some crankshaft angle θ is this the effective mass multiplied by the linear acceleration at the gudgeon pin. Thus:

$$F_I = -m_p\omega^2 r cos\theta\left(cos\theta + \frac{cos2\theta}{n}\right)$$
$$= -m_p\omega^2 r cos\theta - m_p\omega^2 r\frac{cos2\theta}{n}$$

The first term in the above equation is the primary inertia force, while the second term, involving the $cos^2\theta$ term, is called a secondary inertia force. Note that the primary forces change according to the crank angle θ but the secondary forces change at 2θ. Thus, to cancel the secondary forces, any balance shaft that might be designed has to counter-rotate at 2ω.

As well as the inertia forces acting on the reciprocating parts, the forces associated with intake and exhaust pumping, and the pressure arising from combustion of the fuel, also act on the piston and gudgeon pin.

The piston effort is therefore the net force transmitted to the gudgeon pin which are the gas pressure forces (positive or negative), plus or minus the inertia forces (depending on whether the piston is on an upstroke or a downstroke.

2.1.3 Indicated Power

The basis of calculating the performance of heat engines is the area enclosed by a pressure-volume diagram as that shown in Fig. 2.1.4. A piston in a cylinder may compress the gas enclosed in the cylinder (decrease in volume, increase in pressure), or may be forced downwards as a result of heat input (increase in volume, decrease in pressure).

Pressure multiplied by volume has units of energy.

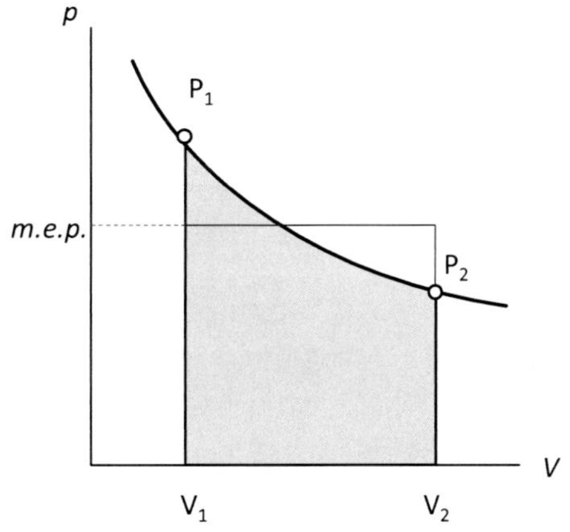

Fig. 2.1.4 *Pressure-volume curve for a gas.*

When a gas goes from one condition (P_1, V_1) to another (P_2, V_2), by either compressing the gas, or the gas expanding, work is done on or by the gas and the amount of work is the area under the pressure-volume curve. When the relationship between pressure and volume is not linear, such as shown in Fig. 2.1.4, the area under the curve can be calculated by integration.

$$W = \int_{V_1}^{V_2} pV\, dV$$

Once the area is known, a straight line at constant pressure can be drawn on the graph which gives the same area as under the curve. The pressure indicated by this straight line is called the mean effective pressure, or m.e.p.

A heat engine is one which involves a series of interconnected *p-V* segments such that a continuous output of mechanical work *W* is produced. It is a consequence of this continuous output that there is always some heat or energy rejected. For a four stroke engine, the theoretical *p-V* diagram is shown in Fig. 2.1.5.

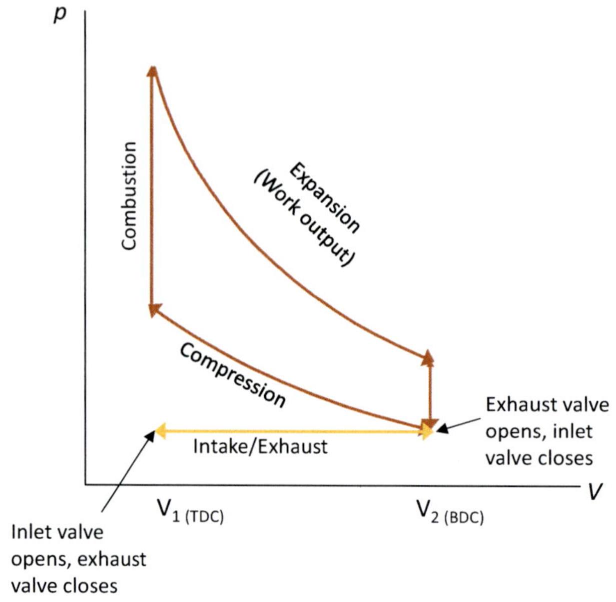

Fig. 2.1.5 *Theoretical p-V diagram for a four stroke engine.*

The p-V diagram shown in Fig. 2.1.5 comprises two cycles. The main working cycle is interrupted by a pumping cycle as exhausted gases are ejected, and a fresh charge of fuel and air is drawn in.

In practice, various losses arising from pumping the gas during intake, exhaust and compression arises, as well as friction losses, so a real p-V diagram is usually measured experimentally with an apparatus known as an engine indicator. This is a mechanical device which charts the stroke of the piston on the horizontal axis (so indicating swept volume) and a connected pressure gauge. The work done, or more commonly, the power measured, is called the "indicated power" of the engine. The indicated power is found by multiplying the work W (area enclosed by the *p-V* diagram from the indicator) by the number of working cycles per second (number of cylinders x rpm/60/2).

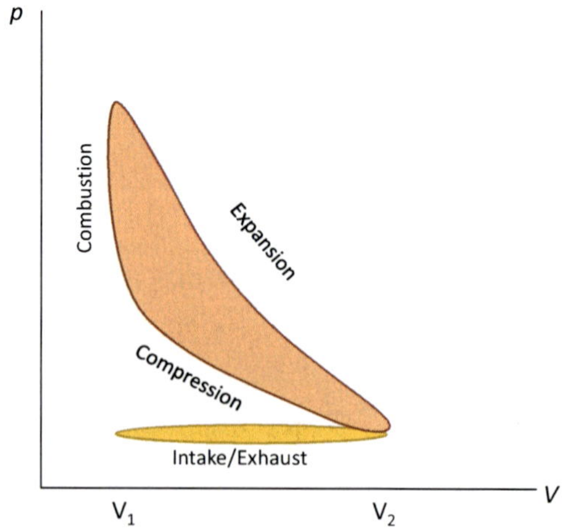

Fig. 2.1.6 *Actual p-V diagram for a four stroke engine.*

36 THE BMC ENGINEERING COMPANION

The area enclosed by the main working cycle is the net mechanical work output, while that enclosed by the pumping cycle represents a mechanical loss. Similar in concept to the mean effective pressure, the brake mean effective pressure is that constant pressure that would be applied to the piston on the power stroke that gives the same energy output, that is, the area enclosed by the p-V diagram. The values for b.m.e.p. are often to be found on an engine performance graph such as that shown in Fig. 2.8.1.

2.1.4 Brake Power

When the output of an engine is to be measured, it is operated under load in a dynamometer. The dynamometer applies a measurable load to the output of the engine. The load can be in the form of a friction brake, a water brake, or even an eddy current electrical brake. The important point is that the engine is braked in some manner.

Fig. 2.1.7 *B series engine on engine dynamometer.*

The measured output power from the dynamometer is thus termed the brake power. Most often, the values are given in horsepower, hence the term "bhp". 1hp = 0.746 kW.

2.2 Efficiency

2.2.1 Thermal Efficiency

Thermal efficiency is the percentage of mechanical work produced as an output relative to the energy input (in this case, the energy released by combustion of the fuel, which is specified by the fuel's calorific value). The calorific value of petrol is about 45 MJ/kg.

An equation for the maximum possible thermal efficiency of a heat engine was formulated by the French engineer S. Carnot. The maximum theoretical thermal efficiency η_i of a heat engine such as that using the Otto cycle is given by:

$$\eta_i = 1 - \frac{1}{\left(\frac{V_1}{V_2}\right)^{\gamma-1}}$$

The adiabatic index γ is the ratio of the specific heats at constant volume and constant pressure for the working substance in the engine. For air, $\gamma = 1.4$. The equation states that other things being equal, the only way to raise the inherent basic thermal efficiency of an engine is to raise the compression ratio.

For a compression ratio of 8.2:1, and assuming air as the working substance, the maximum possible efficiency (with no mechanical friction, thermal, or volumetric losses) is an astounding 56%.

In practice, much energy is wasted: about 30% into the cooling system; 25% as hot exhaust gas; 20% or so by heat loss by convection, conduction and radiation from the engine. By the time all these losses are accounted for, thermal efficiency drops to about 25%. It is considered fortunate if 20% overall efficiency is obtained from an internal combustion engine.

In practice, the overall efficiency can be measured by forming the ratio of the brake power (that measured as being developed by the engine) over the energy input power, the latter being the mass rate of flow of the fuel multiplied by its calorific value.

Diesel engines, with compression ratios of about 18:1, are more fuel efficient than petrol engines and give a theoretical maximum efficiency of about 68%.

2.2.2 Octane Rating

In the 1950s, when most BMC engines were designed, the average compression ratio of most engines of the day was about 7.5:1. This was mainly due to the standard of petrol produced by oil companies. Over the years, petrol became more refined and the compression ratios slowly increased as a result.

The octane rating of a fuel is a measure of its knock resistance when used in an internal combustion engine. Iso-octane is a hydrocarbon fuel that can withstand the pressures of high compression ratios without knocking (or "pinking"). It is arbitrarily assigned an octane rating of 100. Heptane is another hydrocarbon fuel but knocks very easily. It is assigned an octane rating of zero.

To assign an octane number to a particular batch of petrol, the petrol is used in a special test engine which has a variable compression ratio. The compression ratio is increased until knocking occurs. This compression ratio is then noted. The test engine is then kept at this compression ratio and fed with a mixture of octane and heptane. The % iso-octane to heptane is varied until knocking occurs. If, for example, knocking occurs at 80% iso-octane to 20% heptane, the petrol is given an Octane Number of 80. This is called the Research Octane Number (or RON).

Considering that the octane ratio of present day fuels is in the range 90 to 100, it would appear feasible to take advantage of these advances and raise the compression ratio of older engines.

Another other factor that comes into play is the ignition timing. It is desirable to have the maximum combustion pressure occur when the connecting rod has the greatest downwards leverage in relation to the crank pin. To obtain this condition, it is necessary to begin combustion of the fuel before the piston reaches TDC on the compression stroke. For the most part, the static ignition timing is somewhat retarded and limited by the onset of knocking. Technically, knocking occurs due to too high a rate of compression of the mixture by the flame front arising from ignition from the spark plug and should not be confused with pre-ignition arising from hot spots in the combustion chamber or dieseling. Both processes cause rough running and undesirable mechanical loads on the engine.

If 98 octane fuel is used, the ignition timing could probably be increased a few degrees resulting in tangible benefits for almost no cost outlay except for the higher cost of the fuel.

2.2.3 Volumetric Efficiency

Thermal efficiency is concerned with the extraction of mechanical work from the calorific value, or energy content, of the fuel. The theoretical maximum thermal efficiency has been shown to be dependent on the compression ratio.

Various other inefficiencies arise as a result of the basic design of an internal combustion engine, and one of the most important are losses from volumetric inefficiencies.

The ratio of the mass of the incoming charge actually drawn into the cylinder on the intake stroke, compared to the mass of charge that could be drawn in under ideal conditions at atmospheric temperature and pressure is called the volumetric efficiency.

On the exhaust stroke, the piston pushes the products of combustion through the exhaust valve, but cannot displace that volume left in the clearance volume. The gas in this volume is necessarily hot on account of the heat of combustion. The density of the incoming charge is thus reduced by the presence of this hot gas.

Volumetric efficiency is slightly affected by mixture strength. For stoichiometric mixture, and slightly lean, it is a minimum, while for a rich mixture, it is a maximum. The nature of the fuel often plays a significant part in volumetric efficiency since the latent heat of vaporisation means that for some fuels, notably alcohol blended fuels, the volumetric efficiency is increased due to the cooling associated with the transition of the liquid fuel to a vapour. Experiments show that the volumetric efficiency depends on the inverse of the square root of the temperature of the incoming charge.

Volumetric efficiency is decreased with increasing engine speed, and the design of the camshaft is such that volumetric efficiency is maximised for the engine speed range desired. Thus, a high lift cam leads to higher volumetric efficiency at higher engine speeds at the detriment of that at lower engine speeds.

Restrictions in the flow of the incoming charge through the valve decrease the amount (by mass) of combustible mixture drawn in on the intake stroke. These have a greater effect as the engine speed rises and the throttle is opened. The gas velocity at the inlet valve is proportional to the piston area and inversely proportional to the inlet valve throat area. A small cross sectional area of the inlet duct results in a quicker response and good low speed torque, while a large cross section area results in improved volumetric efficiency at higher engine speeds.

When the inlet valve opens, there is a negative pressure pulse generated which travels out towards the atmosphere. When the pulse reaches the atmosphere (at the air cleaner outlet), it is reflected as a positive pulse back towards the valve. If the inlet tract length is tuned to take advantage of the arrival of the positive pressure reflection, a ram flow effect is created which increases volumetric efficiency. Some manufacturers have variable length inlet manifolds to take advantage of this effect. There is an empirical formula which provides an estimate of the optimum length (m) as a function of engine rpm for a single cylinder:

$$L = \frac{2515.24}{rpm}$$

The amplitude of the pressure pulse depends on the square of the engine speed and so has a greater effect at higher rpm.

For a multi-cylinder engine, interference from the opening of other inlet valves may reduce the potential gain via a ram effect.

A casual look the shape of different inlet manifolds that have appeared over the years reflects the change in thinking of the effect of manifold shape on the overall efficiency of the engine. Often as not, increased turbulence of the incoming charge increases power output, but at reduced volumetric efficiency. A log type inlet manifold with square ends imparts turbulence, which helps vaporise the incoming charge, while smooth corners promote unrestricted gas flow thus increasing volumetric efficiency.

The shape of the inlet port is such that it usually has a large entry diameter, but then narrows before reaching the valve so as to increase the gas velocity and promote gas flow and swirl into the cylinder thus introducing turbulence as late as possible.

The volumetric efficiency is related to the mean gas velocity through the input port. Experiments show that the optimum volumetric efficiency at maximum torque is obtained when the mean gas velocity is 43 m/s. Slower than this, the cylinder volume increases faster than the incoming charge enters. Faster than this, turbulence has to greater effect on flow. The ideal valve throat diameter for optimum flow can be calculated from the mean piston speed, V_m, the engine speed N, and the bore B and stroke S.

The mean piston speed is the total distance the piston travels in one revolution, that is, twice the stroke. Thus, for N rpm, the mean piston speed in m/s is expressed:

$$V_m = \frac{(2S \times 10^{-3})N}{60}$$

The piston area is computed from the bore B. The mean volume flow rate Q in m^3/s is thus the piston area times the mean piston speed:

$$Q = V_m \frac{\pi B^2}{4}$$

The same volume flow rate is to be passed through the valve of diameter *d*. Thus,

$$43\left(\pi\frac{d^2}{4}\right) = V_m\frac{\pi B^2}{4} = \frac{(2S \times 10^{-3})N}{60}\left(\frac{\pi B^2}{4}\right)$$

Adjusting for changes from m to mm, the optimum valve throat diameter is shown to be:

$$d = \sqrt{\frac{SND^2}{1290000}}$$

where *N* is the rpm, and all distances *S*, *N* and *D* are in mm.

Usually, the valve opening and closing positions for the inlet and exhaust valves must overlap. This allows some time for the full flow area around each valve to be established, and it is the design of the camshaft that reflects this requirement.

Volumetric efficiency is usually of the order of 80% at full throttle. In practice, motor vehicles are seldom driven at full throttle, and so at part throttle, the volumetric efficiency during normal motoring is quite low.

2.2.4 Mechanical Efficiency

Mechanical efficiency is the ratio of the power actually delivered by the engine to the power developed inside the cylinders as a result of actual combustion.

More formally, it is the ratio of the brake power (as measured by a brake on an engine dynamometer) to the indicated power (as indicated by the enclosed area on a pressure/volume graph).

Mechanical efficiency is usually of the order of 80% and losses arise mainly due to friction (piston rings, bearings, etc), the viscosity of the engine oil, gearbox friction, fan, water pump and charging systems, power steering, etc. Pumping losses associated with intake and exhaust are also causes of loss of mechanical efficiency.

2.3 Combustion Chamber

The influence on efficiency on the compression ratio, and its connection with the octane rating of the fuel, brings into focus the importance of the combustion chamber volume.

In practice, the combustion chamber volume can be measured with a pipette. But the combustion chamber volume is not the compressed volume – that is, the volume of the space above the piston at TDC.

The compression ratio is found from the total volume above the piston at bottom dead centre divided by the total volume above the piston at top dead centre. If we take the swept volume (SV) to be that swept out by the piston from BDC to TDC, and if CV is the compressed volume, then:

$$C.R. = \frac{SV + CV}{CV}$$

CV is the compressed volume. The swept volume *SV* is:

$$SV = \pi\left(\frac{1}{2}b\right)^2 s$$

For a BMC A series 1098cc engine with bore *b* = 64.58 mm and stroke *s* = 83.72 mm, the total swept volume is 1097.2 cc. The compression ratio is specified 8.3:1, and so the compressed volume is thus:

$$CV = \frac{SV}{CR}\left(\frac{1}{1-\frac{1}{CR}}\right)$$
$$= \frac{1097.2}{8.3}\left(\frac{1}{1-\frac{1}{8.3}}\right)\frac{1}{4} = 36.6cc$$

For an Austin 1800 B series engine, the bore *b* is 80.264mm) and *s* = 88.9mm. The compression ratio is 8.2:1 giving a compressed volume of 61.2cc. For engines of interest, calculations and specifications show:

	850	**998**	**998cc Matic**	**1098**	**1275**	**1798**	Unit	**Notes**
Bore	62.94	64.59	64.59	64.59	70.60	80.26	mm	Specified Eng Data
Stroke	68.26	76.2	76.2	83.72	81.33	88.9	mm	Specified Eng Data
Swept Volume (SV)	212.4	249.7	249.7	274.3	318.4	449.8	cc	Calculated per Cyl
Piston Volume (PV)	0.87	5.69	5.69	6.5	6.692	11	cc	Specified on Drawings
Combustion Chamber Face Volume (CCFV)	23.9	23.9	23.9	24.6	31.8	37.9	sq cm	Specified SLS 94 1966
Combustion Chamber Volume (CCV)	24.5	**24.5**	**22.4**	26.1	21.4	38	cc	Specified SLS81 C54/67
Compression Ratio (CR)	8.3	8.3	9	8.5	9.75	8.2		Specified CR
Compressed Volume (CV)	29.09	34.20	31.21	36.57	36.39	62.47		Calculated from CV and SV

Table 2.3.1 *Compression ratio and combustion chamber volumes.*

An interesting question arises. How much can one take off the head to raise the compression ratio? Removing a thickness from the bottom surface of the head affects the value of compressed volume CV in the formula above by virtue of a change to the combustion chamber volume CCV.

Fig. 2.3.1 shows an estimated relationship based upon the data in Table 2.3.1 with a reduction in the CV for a range of thickness removed from the head. Interestingly, service literature[3] quotes a removal of 0.020" from the standard 998cc cylinder head raises the compression ratio to 9:1 (Mini-Matic), but the calculations show a new compression ratio of 8.57 so evidently the newly introduced Cylbestos cylinder head gasket is thinner than that previously fitted to manual transmission vehicles.

[3] C54/67

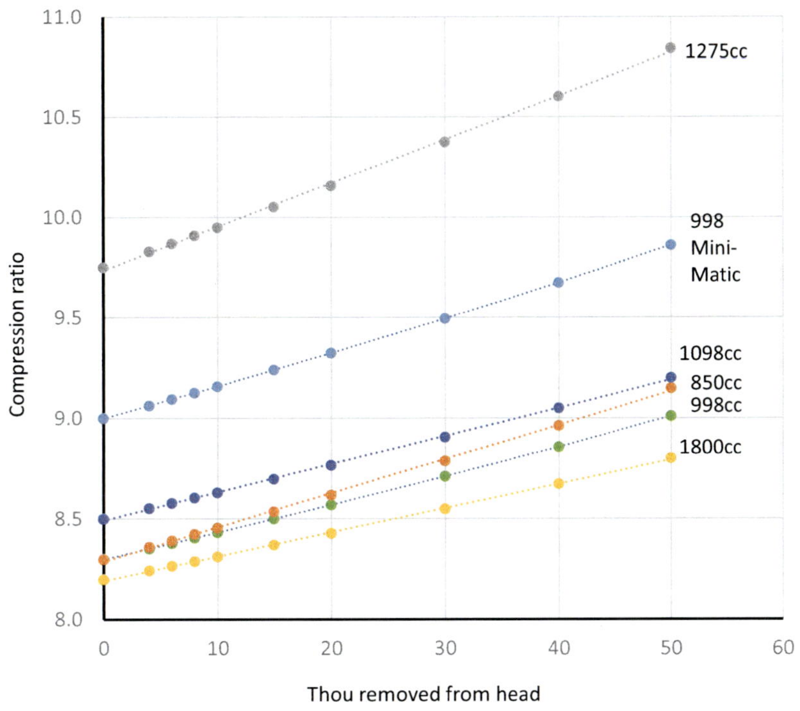

Fig. 2.3.1 *Estimated variation of compression ratio with thickness removed from face of head.*

Given the interplay between the requirements of high thermal efficiency, high volumetric efficiency, and freedom from detonation, the design of a cylinder head is a complicated and specialised affair.

The shape of the combustion chamber is absolutely crucial, and it is no accident that BMC vehicles are adorned with a list of applicable patents on the rocker cover. For BMC engines take advantage of combustion chamber design of Harry Weslake, whose heat-shaped combustion chamber with flat squish area is a familiar sight.

For high thermal efficiency, a combustion chamber and low incidence of detonation, a combustion chamber must have a low surface to volume ratio, result in a short flame travel, and provide turbulence. For a high volumetric efficiency, a high lift camshaft, valve overlap, and large inlet valve diameter is required.

2.4 Engine Vibrations

2.4.1 Turning Moment

Most of the engines fitted to BMC cars are four cylinders. By standards of the day, engine vibration could be expected to be much more pronounced compared to six and eight cylinder cars. Refinements in balancing and engine mounting makes modern four cylinder cars very much smoother and comparable to sixes of yesteryear. So, what exactly makes a four cylinder engine so rough and what is done by the manufacturer to bring vibrations under control for the comfort of the occupants and reliability of the mechanicals?

Because these engines use a four stroke Otto cycle, the power stroke for one cylinder occurs on every second revolution of the crankshaft – that is, the delivery of power to the crankshaft is not continuous like a turbine engine but is intermittent. Crankshafts are arranged so that two pistons are "up" while the other two are "down". The crankshaft is in a single plane. Of the pistons that are "up", one is on the exhaust stroke while the other is on the compression stroke. Of the pistons that are "down", one is on the power stroke and the other is on the inlet stroke. So, the crankshaft receives regular pulses two per revolution from different cylinders. By the time the crankshaft has rotated two revolutions, each piston has provided one power stroke. The flywheel (on a manual car) serves to dampen out variations in crankshaft speed by alternately storing rotational energy and then releasing it as the crankshaft attempts to speed-up and slow-down in between power stroke delivery.

These concepts are often described graphically in the form of a turning moment diagram. For one cylinder, a somewhat simplified turning moment diagram is shown as the resultant of the positive and negative moments arising from compression, combustion and inertia loads as shown in Fig. 2.4.1.

The sum of the areas above the zero line less the sum of the areas below the zero line represents the net energy available as output. This energy, divided by 4π gives the position of the mean turning moment, or torque, on the diagram. The intersection of the resultant turning moment with the mean gives the points of maximum and minimum angular speeds of the crankshaft. These correspond to points of maximum kinetic energy and minimum kinetic energy of the rotating mass. That is, when the resultant is above the mean, the crankshaft is speeding up, and when below the mean, the crankshaft is slowing down. The greatest fluctuation of energy occurs between points of minimum and maximum rotational speeds.

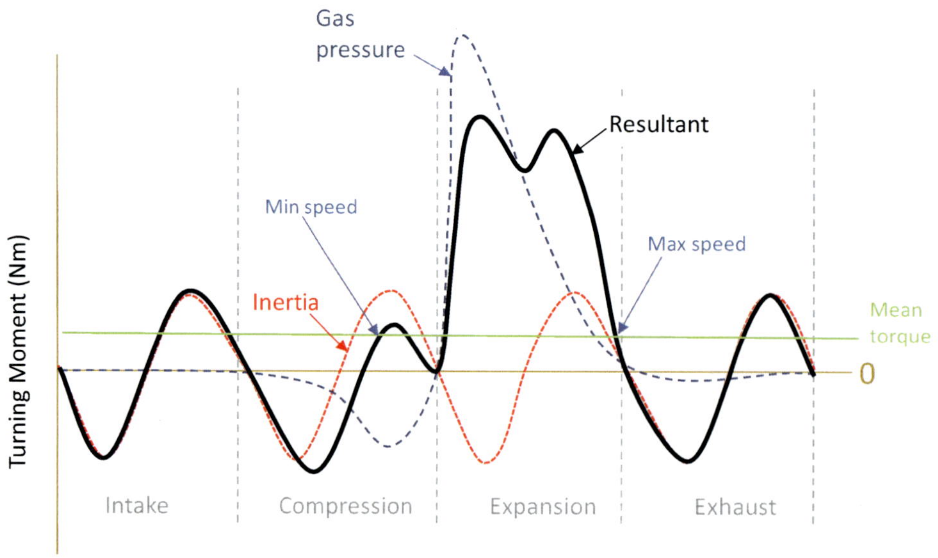

Fig. 2.4.1 *Turning moment diagram single cylinder engine.*

For a multi-cylinder engine, the turning moment diagrams for each cylinder are simply added together.

Expressed as a ratio of angular velocities ω, the coefficient of fluctuation of speed K_s is given by:

$$K_s = \frac{\omega_{max} - \omega_{min}}{\omega}$$

where ω is the mean angular velocity. For an automotive application, K_s is designed to be about 0.03.

The coefficient of fluctuation of energy K_E is the ratio of the maximum fluctuation of energy *K* per cycle to the mean kinetic energy *E* of the rotating parts.

$$K_E = \frac{K}{E}$$

For a four cylinder engine, K_E is usually designed to be about 0.3 to 0.4.

These design parameters are met by the use of a flywheel. The size of the flywheel is expressed as its moment of inertia *I*. Thus,

In general, the energy in a flywheel of moment of inertia *I* rotating at angular velocity ω is:

$$K = \frac{1}{2} I \omega^2$$

And so the maximum fluctuation in the flywheel on a reciprocating engine is thus:

$$K = \frac{1}{2} I \left(\omega_{max}^2 - \omega_{min}^2\right)$$

The coefficient of fluctuation of energy is:

$$\begin{aligned} K_E &= \frac{1}{2} I \frac{\left(\omega_{max}^2 - \omega_{min}^2\right)}{E} \\ &= \frac{1}{2} I \frac{(\omega_{max} + \omega_{min})(\omega_{max} - \omega_{min})}{E} \end{aligned}$$

Since the speed variation is usually small,

$$\omega_{max} + \omega_{min} \approx 2\omega$$

And so,

$$K_E = \frac{1}{2} I \frac{(2\omega)(\omega_{max} - \omega_{min})}{E}$$

The coefficient of fluctuation of speed is expressed:

$$K_S = \frac{\omega_{max} - \omega_{min}}{\omega}$$

$$\omega_{max} - \omega_{min} = K_S \omega$$

And so the coefficient of fluctuation of energy K_E can be written:

$$= \frac{1}{2} I \frac{(2\omega) K_S \omega}{E}$$

$$I = \frac{K}{K_S\omega^2}$$

K_E and E are functions of engine layout (e.g. number of cylinders). K_S depends on what is tolerable for the application. Note that because there is an ω^2 term in the denominator for the expression for I, the greater the operating speed, the smaller the flywheel required by a considerable amount.

The moment of inertia I of a flywheel depends upon the distribution of mass within it. For example, for a solid disc of radius R and mass m, I with respect to rotation about the axis is given by:

$$I_C = \frac{1}{2}mR^2$$

For a hollow cylinder of outer radius R and inner radius r, with the mass concentrated at the rim, the moment of inertia is given by:

$$I_C = \frac{1}{2}m(R^2 + r^2)$$

The effect of this geometry is that the mass, concentrated at the rim, makes a larger contribution to the moment of inertia of the part compared to the case where the same mass is uniformly distributed throughout.

The moment of inertia of the flywheel affects the computation of what is called the effective mass of the vehicle when determining the tractive effort available for acceleration. This important issue will be discussed in Section 2.8.

2.4.2 Balance

With reciprocating piston engines, the primary force associated with the reciprocating motion of the pistons and connecting rods is the main cause of imbalance. In a single piston engine (such as in a lawn mower), the piston is acting like a stone being whirled around on a length of string. Mass is usually added to the crankshaft webs to counterbalance the reciprocating mass of the piston, but this can only provide a certain degree of balance since the effect is such that whatever piston mass is balanced in the vertical direction, additional out of balance mass is found to exist in the horizontal direction. In a 90° V twin engine, additional of a crankshaft balance mass would therefore be very effective.

In an in-line multiple cylinder engine, one may initially think that the force on the pistons flying upwards would be balanced by the forces on the pistons flying downwards and so perfect balance would be achieved. This is true. The primary forces are balanced – but only for the masses of the parts, this does not take into consideration the impulses arising from combustion within the head. The revolving mass of each crank pin (which acts through 360°) leads to a couple, or a bending of the crankshaft and so counterweights are attached on the opposite side of the crankpin to balance these rotating masses.

So, if the primary and rotating masses are balanced by counterweights, why is a four cylinder engine not as smooth as a 6 or 8 cylinder? What is not generally appreciated is that there is an inherent difference between the force necessary to bring a piston over top dead centre compared to going through bottom dead centre. This is because the swing of the connecting rod acts against the swing of the crankshaft at TDC while it acts with the rotation of the crankshaft at BDC. At TDC, the piston is going over a hill, while at BDC, the piston is going down into a valley. This difference in motion manifests itself as what are called secondary forces – the shorter the connecting rod, the larger the effect. In a four cylinder engine, it is impossible to arrange the crankshaft and firing order so that these secondary vibrations cancel each other out with even firing pulses – but in a six cylinder engine, this can be done. This is why a six cylinder engine is so smooth. A six cylinder engine is inherently balanced for both primary and secondary forces. In an eight cylinder engine, secondary vibrations cannot be cancelled completely, but can be minimised more so than in a four cylinder. The Chrysler/Mitsubishi Sigma "silent shaft" engine was designed to reduce the secondary forces to a minimum and it was indeed a smooth four cylinder engine.

2.4.3 Engine Mountings

Designers employ several methods to make the engines acceptably smooth – the most obvious being the addition of a flywheel. Not so obvious is the harmonic balancer (usually attached or part of the crankshaft pulley). The harmonic balancer is used to dampen torsional vibrations of the crankshaft. This pulley is in two parts: An outer heavy rim, and an inner lighter disc which is keyed to the crankshaft. The inner part is separated from the outer rim by a rubber bonded ring. Sometimes, this rubber extrudes out of its housing and the pulley has to be replaced. This often happens if the crankshaft pulley becomes soaked in engine oil due to a faulty or worn timing cover seal, or is overloaded by having to drive air conditioning and power steering loads. What is not often detected is that the rubber can also harden with age and become somewhat petrified. In this case, it is amazing the amount of vibration that can be eliminated by simple changing the pulley to a new one with pliable rubber instead of "stone".

The next most obvious remedy to reduce vibrations being transmitted to the body is to mount the engine on flexible mountings. A cured rubber is the usual medium. The objective here is that the vibrations are absorbed by the rubber before they are transmitted to the body structure. The location of the engine mounts is very important. As well as absorbing vibrations, the mountings also have to restrain the engine's reverse reaction torque, support the weight of the engine, and to also restrain the engine from becoming a "loose cannon" during heavy braking – that is, acting as a seat belt for the engine.

The location and number of engine mounts is very important. For the sake of minimum vibration, the engine mounts need to be placed where the vibration of the engine is at a minimum. That is, if the engine is sitting there, a vibrating lump, then the minimum vibration occurs along an axis through its centre of gravity. More correctly, the front mountings should be placed along a line that joins the centre of mass to the centre of percussion (a point much like the sweet spot in a tennis racquet where no shock is felt though your arm) relative to the rear engine mount. The centre of percussion differs from that of the centre of gravity in the sense that it takes into account the rotational of the distribution of mass rather than the static distribution.

In the Austin 1800, this was the intention of the original UK style of mount – two at the front somewhat low down but near the centre of mass as looked at in the vertical direction (i.e. about middle of the engine and gearbox) with the rear single mount along the transverse centreline.

Australian engineers found that the UK style of rear mount for this vehicle failed due to the heat from the engine block affecting the rubber to metal bond. Initially, an insulating asbestos spacer was made available, and then a short time later, the "outrigger" style mountings were devised.

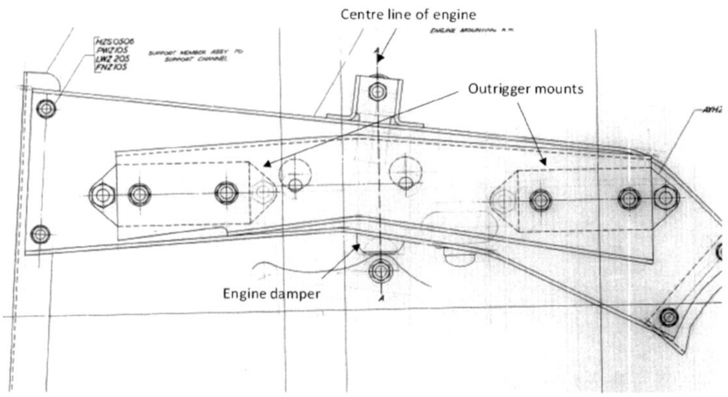

Fig. 2.4.2 *Outrigger engine mounts, Austin 1800, ADO17.*

Although the outrigger style mountings removed the rubber parts out away from the block, they also moved the mountings out from the centre line to almost the same distance as the front mountings and so any vibrations are magnified by the resulting leverage as a result. However, placing engine mountings at the centreline also means that the capacity for taking care of reverse torque reaction from the engine is also very much diminished and this is why we have widely space engine mounts at the front of the block as well as a tie rod at the rear – which on Australian Austin 1800 cars had to be moved to the underneath of the gearbox because the original position (up near the rear engine mount was no longer available due to the outrigger mounting).

Another notable feature of the Austin 1800 engine mountings is the inclusion of a damper, or shock absorber, connected between the engine centre line at the bell housing and the engine mounting cross member. Evidently balance of this power unit was difficult to achieve and the damper a costly addition.

2.5 Camshaft

2.5.1 Introduction

One of the most important design features of an automotive engine is that of the camshaft. Apart from the necessity of opening and closing the inlet and exhaust valves in the proper sequence, comes all the second order issues such as mechanical reliability, the effect on volumetric efficiency, valve head cooling, tractability, and the engine rpm at which maximum torque and maximum power become available.

Associated with the camshaft are the followers or tappets, the pushrods, and the rocker arms. The cam profile must be developed to impart a suitable motion to all these parts for a wide range of engine rpm reliably since any failure will most likely result in extensive damage to the rest of the engine.

The camshaft turns at half engine speed.

2.5.2 Follower Motion

The camshaft converts rotary motion geared down from the crankshaft into a reciprocating motion of the follower. The cam profile depends upon the desired motion of the follower.

2.5.2.1 Constant Velocity

A simple cam might provide a uniform velocity to the follower on the upstroke and a uniform velocity on the downstroke. Except where the velocity changes direction, the acceleration is thus zero throughout. The follower displacement, that is, the lift, is thus:

$$s = \frac{v}{t} = \frac{\theta}{\omega} v$$

where v is a constant. The displacement, velocity and acceleration as a function of θ are shown in Fig. 2.5.1 (a).

2.5.2.2 Constant Acceleration

Another cam profile might provide constant acceleration of the follower on the first half of the upstroke and a constant deceleration during the second half of the upstroke, followed by the reverse on the downstroke. This profile is in the form of an eccentric such as that which might be used to operate a fuel pump.

Let a cam rotate at angular speed ω. The requirements are that for the first part of the follower motion, there is a constant acceleration, followed by a constant deceleration as the follower reaches full lift. The angle θ represents the angular displacement of the cam.

The distance s of the follower varies in some way which is not immediately apparent. During a small increment of angle $d\theta$, the follower moves a distance ds with a change in velocity dv. There must be a change in velocity because our initial conditions are that the accelerations of the follower are constant. If the acceleration during the rotation $\theta = 0$ to θ_1 is a constant, then

$$\frac{dv}{d\theta} = constant$$

If $dv/d\theta$ is a constant, then a graph of v vs θ from $\theta = 0$ to θ_1 is a straight line.

With reference to the velocity v,

$$\omega = \frac{d\theta}{dt}$$

$$\frac{dv}{dt} = a_1$$

$$= \frac{dv}{d\theta}\frac{d\theta}{dt}$$

$$dv = \frac{a}{\omega} d\theta$$

$$\int_0^{v_1} dv = \frac{a}{\omega}\int_0^{\theta_1} d\theta$$

$$v_1 = \frac{a}{\omega}\theta_1$$

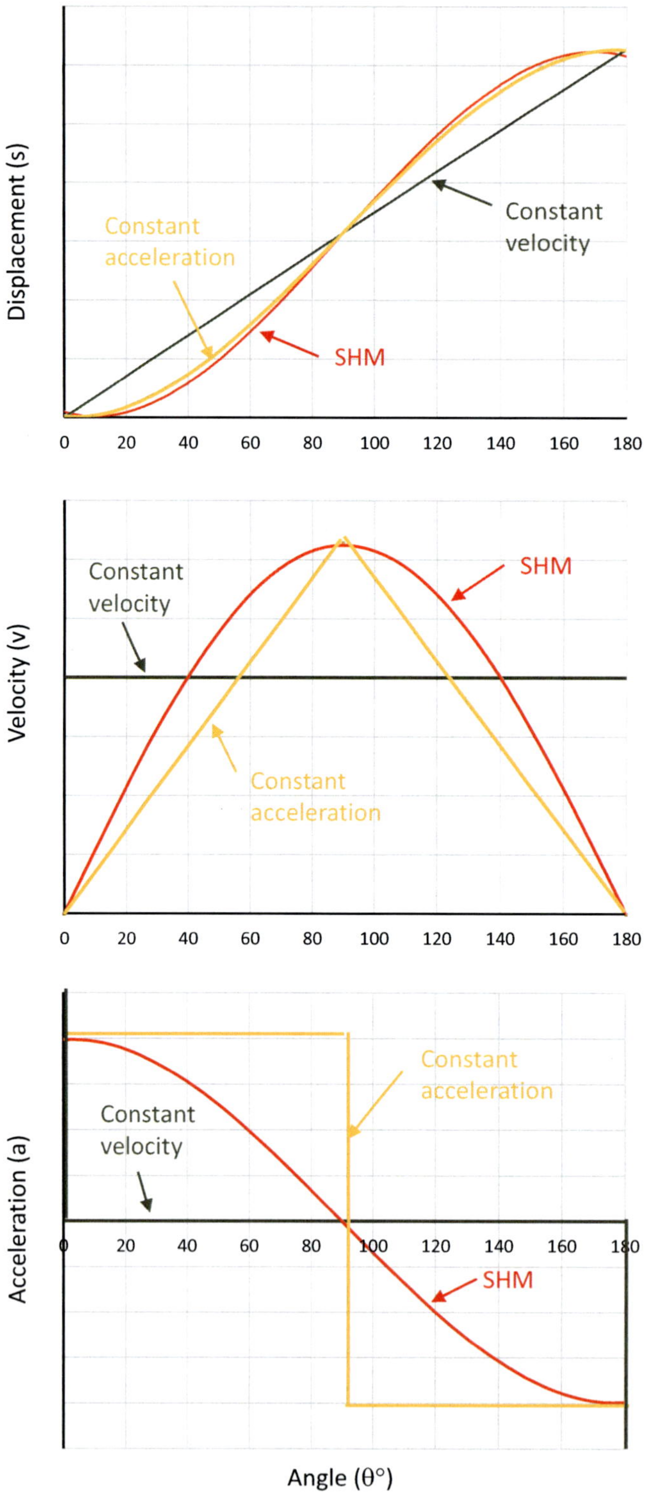

Fig. 2.5.1 *Displacement, velocity and acceleration profiles for various cam profiles.*

Since a and ω are constant, then there is a linear relationship between v and θ and so in general,

$$v = \frac{a}{\omega}\theta$$

With reference to the acceleration a,

$$\begin{aligned} v &= \frac{ds}{dt} \\ &= \frac{ds}{d\theta}\omega \\ ds &= \frac{v}{\omega}d\theta \\ \int_0^{s_1} ds &= \frac{1}{\omega}\int_0^{\theta_1} v(\theta)d\theta \\ s_1 &= \frac{a}{\omega^s}\frac{\theta_1^2}{2} \\ a_1 &= \frac{2s_1\omega^2}{Q_1^2} \end{aligned}$$

Similar expressions can be obtained for the angular displacement from θ_1 to θ_2.

If the total displacement s is $s_1 + s_2$, and for the special case of $\theta_1 = \theta_2$ (i.e. the acceleration up to s_1 is equal in magnitude to the deceleration from s_2 back to 0, then the acceleration a as a function of θ can be shown to be:

$$a = \frac{4\omega^2}{\theta^2}s$$

and

$$\begin{aligned} v &= \frac{a}{\omega}\theta \\ s &= \frac{a\theta^2}{4\omega^2} \end{aligned}$$

The displacement, velocity and acceleration of the follower for this example are shown in Fig. 2.5.1. Note that for the case of a constant acceleration, the shape of the displacement against angle θ is a square law.

This geometry results in an instantaneous change of sign of the acceleration, but the overall magnitude of the acceleration is a minimum compared to other simple profiles. This type of geometry can be used for low speed internal combustion engines. During the early part of the lift, and the later part of the close, an external spring load is usually needed to keep the follower on the cam.

2.5.2.3 Simple Harmonic Motion

Another possible cam profile is one which imparts simple harmonic motion to the follower in the sense that the velocity and accelerations are sine and cosine curves respectively. For the sake of a convenient datum, it is convenient to express the displacement s in terms of the maximum displacement s_m as:

$$s = \frac{s_m}{2}(1 - cos\theta)$$

In this equation, it is assumed that the maximum displacement occurs at θ = 180° but this need not be so in practice if there is a flat at the top of the cam where the follower, and valve, is held open at maximum lift for some portion of the camshaft rotation. If the displacement reaches a maximum before half a turn of the crankshaft, then the angle θ has to be scaled by π/θ_m where θ_m is the angle at where s reaches s_m. Thus,

$$s = \frac{s_m}{2}\left(1 - cos\left(\theta\frac{\pi}{\theta_m}\right)\right)$$

Differentiating with respect to θ, and since

$$v = \frac{ds}{d\theta}\frac{d\theta}{dt}$$

and $d\theta/dt = \omega$, the velocity v as a function of θ is thus:

$$v = \frac{\omega}{2}\frac{\pi}{\theta_m}s_m sin\left(\theta\frac{\pi}{\theta_m}\right)$$

The velocity is a maximum when $\pi\theta/\theta_m = \pi/2$.

$$v_m = \frac{\omega}{2}\frac{\pi}{\theta_m}s_m$$

The acceleration is the derivative of the velocity, and so since

$$a = \frac{dv}{d\theta}\frac{d\theta}{dt}$$

then:

$$a = \frac{\omega^2}{2}\frac{\pi^2}{\theta_m^2}s_m\ cos\left(\theta\frac{\pi}{\theta_m}\right)$$

The acceleration is a maximum when $\pi\theta/\theta_m$ = 0 and 2π and so:

$$a_m = \frac{\omega^2}{2}\frac{\pi^2}{\theta_m^2}s_m$$

The displacement, velocity and acceleration of the follower for this example are shown in Fig. 2.5.1.

The maximum acceleration is important in these equations because this determines the inertia loads put on the follower, push rod, valve springs, and valve and so limits the maximum rotational speed of the engine.

2.5.3 Cam Profiles

Fig. 2.5.2 shows the arrangement of cam lobes for a 998cc power unit, Morris Mini Clubman, YDO21.

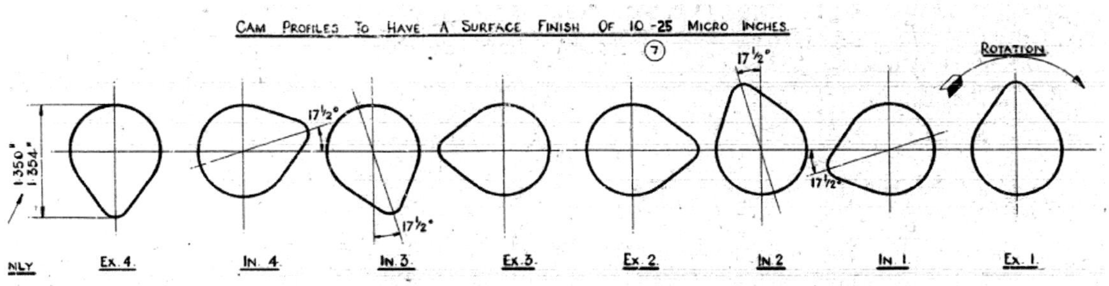

Fig. 2.5.2 *Cam lobe arrangement for 998cc, YDO21.*

The cam profiles for inlet and exhaust reflect the desired driving characteristics. For models where the market can tolerate decreased driveability at low engine speeds, the profiles are such that maximum torque is relatively low, about 2500 rpm. For high performance models, the profiles are selected for a greater lift at higher engine speeds.

The basic profile for the exhaust valve for the above camshaft is shown in Fig. 2.5.3.

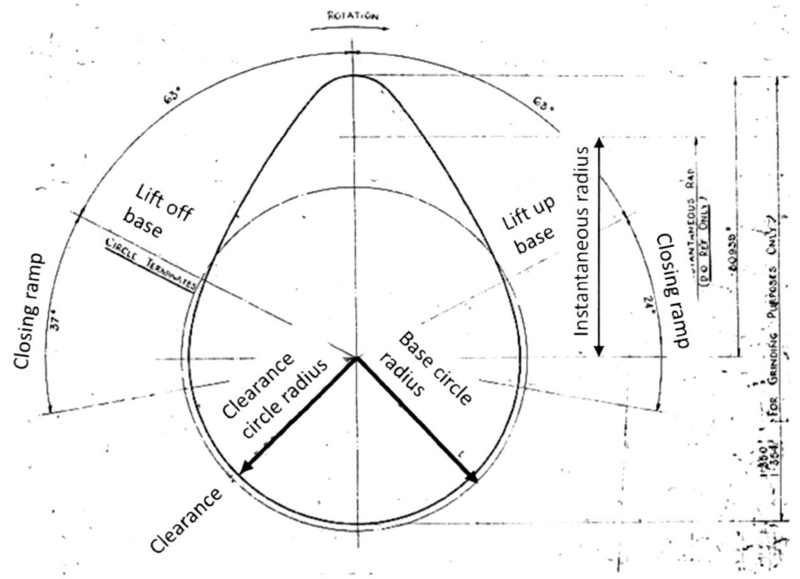

Fig. 2.5.3 *Cam profile basic dimensions for exhaust lobe for 998cc, YDO21.*

Table 2.5.1 shows the differences in basic dimensions of the cam shown in Fig. 2.5.3. Notice that the dimensions are specified to 0.00001".

	Inlet	Exhaust
Base circle radius	0.55933"	0.55933"
Clearance circle radius	0.54257"	0.54261"
Clearance	0.01676"	0.01678"
Closing ramp angle	40°30'	24°
Opening ramp angle	40°30'	37°
Apex half-angle	68°	63°
Instantaneous radius	0.7066"	0.60620

Table 2.5.1 *Basic measurements of inlet and exhaust cam profiles, 998cc, YDO21.*

Apart from the base circle which is easily defined, the shape of the opening and closing ramps, and the nose are usually tabulated in terms of the cam interval in degrees, the tangent lift off the base circle and the tangent height above the cam centreline in 1° steps.

2.6 Clutch

2.6.1 Clutch Performance

The clutch allows disconnection of drive from the engine to the transmission for the purpose of changing gears. The synchromesh cones within the transmission speed up, or slow down the input drive train (including the clutch driven plate), to match the rotation of the output gear train which is determined by the speed of the vehicle. When disengaged, the clutch must transmit full engine torque to the input gear train of the transmission.

For smooth gear changes, the clutch driven plate should have as low as a moment of inertia as possible so as to reduce the load and wear on the synchromesh. Since moment of inertia depends upon the total mass, and how that mass is distributed, it is found that the clutch driven plate is usually a thin metal disc with friction lining around the edges so that the total mass may be reduced, but an adequate level of torque may be applied and transmitted under a reasonable spring pressure with while providing an acceptable pedal effort for the driver.

For a Morris 850, ADO15, the dimensions of the driven plate are shown in Fig. 2.6.1 and the arrangement of the driven and driving plates and springs in Fig. 2.6.2. The clutch for other models in the Mini range is similar with the exception of the spring unit, which in later models is of a diaphragm type in the manner of a Belville washer with a 12° included angle.

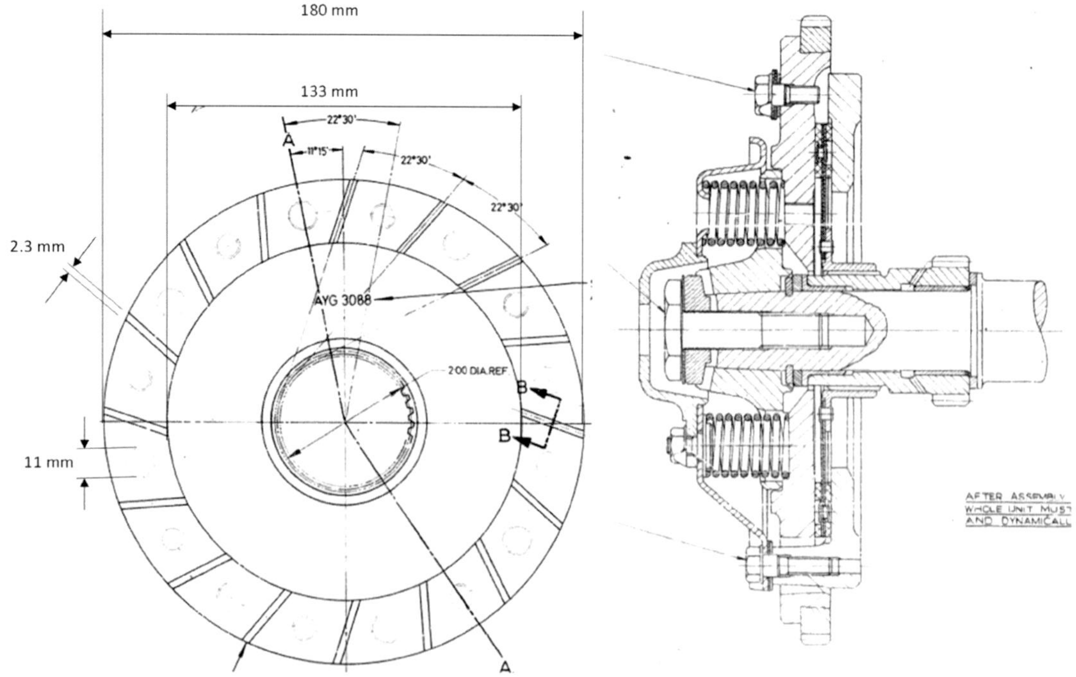

Fig. 2.6.1 *Clutch driven plate Mini range (B354) and arrangement.*

The torque capacity of the clutch can be readily determined by its dimensions, the coefficient of friction (usually about 0.3), and the spring pressure expressed as a force. The maximum friction force F is found from:

$$F = \mu N$$

N is the clamping force produced by the springs. In the case of the arrangement shown in Fig. 2.6.2, each of the six springs has a rating of 65 lb or 289 N. The total transmissible torque is then the friction force multiplied by the mean radius of the facings. Table 2.6.1 shows the results for Morris 850 (ADO15) and Morris Cooper S (YDO16), the latter fitted with a diaphragm spring rated at 2898 N.[4]

	850cc	1275cc	Unit
OD	180	180	mm
ID	133	133	mm
Coefficient of Friction	0.3	0.3	
Spring force spec	289.12	2898	N per spring
No. springs	6	1	
Total clamping force	1734.72	2898	N
Average radius	78.25	78.25	
Friction force (max) per side	520.4	869.4	N
Total torque (2 sides)	81.4	136.1	Nm
Engine Torque (Max)	59.6	106.9	Nm
Overrated by	36.7	27.3	%

Table 2.6.1 *Clutch performance calculations 850cc and 1275cc.*

Also shown in Table 2.6.1 are the specified engine maximum torque, and the % over-rating of the clutch. An overrating of about 30-40% is usual in clutch design to allow for manufacturing tolerances and abuse generated from improper operation by inexperienced drivers.

The wear on the clutch driven plate facings is determine by the pressure, which in turn can be calculated from the frictional area of contact. Allowance must be made for the area taken up by the rivets and the grooves in the facings. The pressure at the clutch lining is shown in Table 2.6.2.

	850cc	1275cc	Unit
Angle of grooves	22.5	22.5	deg
Width of grooves	2.3	2.3	mm
Length of grooves	25.4	25.4	mm
No. grooves	16	16	
Area of grooves	936.1	936.1	mm^2
Dia of rivets	11	11	mm
No. Rivets	16	16	
Area of rivets	1520.5	1520.5	mm2
Nominal friction area per side	11554.0	11554.0	mm2

[4] Manufacturer's rating for 6.25" (159mm) dia clutch with 6 coil springs is 45 ftlb (61 Nm) and for an equivalent diaphragm spring clutch of 600 lb (2688 N), 75 ftlb (102 Nm). SLS55.

Actual friction area per side	9097.4	9097.4	mm2
Total clamping force	1734.72	2898	N
No sides	2	2	
Pressure at lining	57.20	95.57	kPa

Table 2.6.2 *Clutch facings pressure calculations 850cc, 1275cc.*

Given the increased spring pressure for the 1275cc engine, it is not surprising to find that the lining material specified for this engine is different to that for the 850cc engine. The increase in rating is also reflected in the pedal pressure computed from the mechanical and hydraulic leverage offered by the throwout arm, slave and master cylinders, and the pedal. Table 2.6.3 shows the computed pedal effort necessary to just move the pressure plate off the clutch plate.

	850cc	1275cc	Unit
Throwout lever			
Long side	154.0	154.0	mm
Short side	27.0	27.0	mm
Offset	38.9	38.9	mm
Effective Long Side	151.6	151.6	mm
Lever ratio	5.7	5.7	
Master and Slave Cylinders			
Slave cylinder force	304.0	507.9	N
Dia of slave cylinder	22.2	22.2	mm
Pressure in clutch line	783.7	1309.2	kPa
Dia of master cylinder	19.1	19.1	mm
Hydraulic ratio	1.4	1.4	
Master cylinder force	223.4	373.1	N
Pedal			
L1	244.5	244.5	mm
L2	58.7	58.7	mm
Lever ratio	4.2	4.2	
Foot pedal effort	53.6	89.6	N
	5.5	9.1	kg

Table 2.6.3 *Computed pedal effort required to engage the clutch.*

It is interesting to note that at the point of disengagement from a standing start, the clutch is required to momentarily dissipate full engine power via friction at the facings.

2.6.2 Clutch Release Bearing

The clutch release bearing is called on to apply at least the clamping force produced by the springs. Later model vehicles with a diaphragm spring clutch are fitted with bearing DAM1654 which has a thrust load rating is 1595 lbf (approximately 7 kN). This bearing is often supplied as a replacement part for all models.

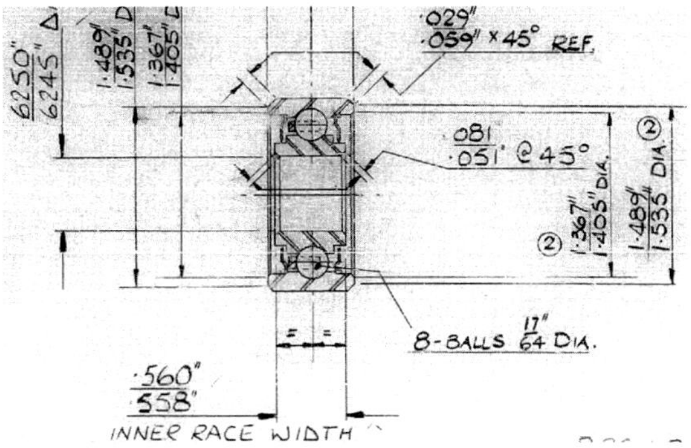

Fig. 2.6.2 *Clutch release bearing DAM1654.*

DAM1654 appears adequate considering the 2.9 kN clamping force required for the 1287cc engine. The bearing is called upon to provide more than the clamping force since the spring has to be compressed further to provide running clearance for the driven plate when the clutch is engaged.

For the coil spring clutch, Table 2.6.4 shows the load at maximum compression for each spring calculated on the basis of the specified free length, fitted length and fitted load, and maximum compression, according to Hooke's Law.

Bearing load - coil spring	850cc	Unit
Free Length	68.12	mm
Fitted length	35.61	mm
Fitted load	289.12	N
Spring constant	8.89	N/mm
Maximm deflection	30.73	
Load at max deflection	332.29	N
Throwout bearing load	1994	N

Table 2.6.4 *Coil spring clutch calculated maximum load for one spring.*

With 6 springs, the maximum loading on the throwout bearing is approximately 2kN. This load is of course then supported by the crankshaft thrust washers. This is well within the capability of the throwout bearing fitted.

When first introduced in Australia in 1961, the clutch release bearing for the Mini range was Part No. 7MJT5/8 which is a Ransom and Marles thrust bearing 5/8" ID (BMC Part No 2A3653). Over the years, this would change to Part Nos are 3/W5/8, CR5/8B, GRB201, GRB238 and DAM1654.

2.7 Transmission

2.7.1 Introduction

Why do we have a gearbox in our cars? These days, and especially with the popularity of automatic transmissions, we tend not to give the question a thought.

Firstly, it is immediately obvious that some means of disconnecting the engine from the drivetrain is required, at least for a manual transmission, for the vehicle to be brought to a halt and yet with the engine running. This is one of the jobs of the clutch.

The engine develops torque, which must in turn be transmitted to the road as tractive effort in order to propel the vehicle forward. The engine, by virtue of its mechanical arrangement, typically turns within a range from about 1000 rpm to about 6000 rpm. It is not practical for the road wheels of the usual size to turn at such a rate, and so an overall reduction in speed, brought about by a reduction gear, is required to match the speeds of the engine with the wheels, while keeping in mind that the diameter of the road wheels is usually within the range of 10 to 14 inches. This reduction is provided by the final drive gear, typically a ratio of about 3 to 4:1.

Finally, speed and road conditions vary considerably throughout a journey. The engine develops torque at a fairly defined band of rotational speed, and maximum power at a greater speed. When the vehicle is required to climb a gradient, more torque is required compared to the case of running on level ground. When starting off from rest, the torque, and hence tractive effort, is much greater than that required at higher speeds so that an acceptable level of acceleration can be attained to reach the desired road speed and once at that speed, the engine is operating efficiently.

The gearbox, or transmission, is thus a means of altering the ratio of engine speed to road wheel speed, and consequently torque, and thence tractive effort, for convenient propulsion of the vehicle.

In a conventional transmission, as distinct from a continuously variable transmission, fixed gear ratios are provided and selection between them made by the driver through means of the clutch.

Within the transmission, there is a mechanism by which gear ratios are selected and engaged without producing undue mechanical shock to the components. In most cases, this mechanism takes the form of a synchromesh cone clutch which speeds up or slows down the input drive train (including the clutch driven plate), to match the rotation of the output gear train which is determined by the speed of the vehicle.

2.7.2 Engine Speed Ratio

To avoid an undue amount of gear changes to be made by the driver, and mechanical wear and tear, it is desirable that the engine be operated at all times at a speed which is at or above that of where it develops maximum torque.

To see why this might be so, imagine that the engine is operated in each gear up to that of maximum power before an up change to the next higher gear is made.

When changing gear from say 2nd gear to 3rd gear, the engine rpm may fall below that at which maximum torque occurs. If there is then an increase in road grade or a slow down due to traffic conditions, the engine speed may drop further and even less torque available, necessitating a down change. The effect may be dramatically illustrated if one attempts to change up from 2nd gear to top gear without going through 3rd gear. In such a circumstance, there may not even be enough engine torque to maintain current speed at all.

If, on the other hand, the up change results in a drop in engine rpm that was above that of maximum torque, then a minor increase in grade will result in a change in engine speed downwards towards that of maximum torque – that is, more engine torque available and reducing the likelihood of a down change.

The ratio of the engine speed at maximum torque to maximum power is called the *engine speed ratio* and this determines the spacing of gear ratios in the transmission.

2.7.3 Gear Ratios

In the simplest case, the gear ratios are arranged in a geometric progression. If s_1, s_2, s_3, and s_4 are the gear ratios for 1st, 2nd, 3rd and 4th speed gears, then for a geometric progression *g*, we have:

$$\frac{s_2}{s_1} = \frac{s_3}{s_2} = \frac{s_4}{s_3} = g$$

The common ratio between each gear is *g*. If 1st gear is given the gear ratio *k*, then $s_1 = k$, $s_2 = kg$, $s_3 = kg^2$ and $s_4 = kg^3$.

If the engine is to operate within the range of maximum power and maximum torque, then:

$$g \geq Engine\ speed\ ratio$$

Top gear is usually a 1:1 ratio through the gear box, and, in most of our BMC vehicles, there are four forward speeds. Thus,

$$kg^3 = 1$$
$$g = \frac{1}{\sqrt[3]{k}}$$

For gearboxes of *n* numbers of intermediate gears, the value for *g* can be computed:

$$g = \frac{1}{\sqrt[n-1]{k}}$$

The number of ratios, *n*, is chosen so that there is enough number of intermediate gears to ensure that the engine will operate at or above maximum torque at all possible road speeds. Thus, the bottom gear ratio *k* is given by:

$$k \leq \frac{1}{g^{n-1}}$$

Let's see how this works for an A series 1098cc engine. In the following discussion, we will compare the characteristics of a 1098cc transmission with a 1275cc transmission.

For a 1098cc engine such as that fitted to 2/YDO5 Morris Mini 1100, maximum power occurs at 5100 rpm and maximum torque at 2500 rpm. Manufacturers select a 1st gear ratio to suit a design specification for gradeability. In the case of a Morris Mini 1100, the 1st, or bottom, gear ratio is 3.53.

Thus, the common gear ratio *g* has to be such that:

$$g \geq \frac{2500}{5100} = 0.49$$

Let us compute the theoretical values of *k* for *n* = 2, 3 and 4 using the limiting value of *g* above.

n	*k*
2	2.040
3	4.162
4	8.490

Table 2.7.1 *Values of bottom gear ratio k for limiting value of g for 2nd, 3rd and 4th gear (1098cc).*

Remembering that *k* is the ratio for 1st gear, then we can see that a 2 speed gearbox would be unsuitable for this application. A 3 speed gearbox would suffice at *k* = 4.165 which is higher than that actually fitted at 3.53, but a 4 speed gearbox gives far greater room for flexibility in the in-between ratios.

For a 1275cc Cooper S power unit, the same calculations, with a manufacturer's 1st gear ratio of 3.2:1 and maximum power at 5800 rpm and maximum torque at 3000 rpm (engine speed ratio 0.52), we obtain:

n	*k*
2	1.933
3	3.738
4	7.226

Table 2.7.2 *Values of bottom gear ratio k for limiting value of g for 2nd, 3rd and 4th gear (1275cc).*

In this application, a 3 speed gearbox is getting close to the bone, whereas a 4 speed gearbox allows more flexibility.

We need this flexibility because gear changes are not instantaneous in practice and during the time for a gear change, particularly between 3rd and 4th which are more frequently made than between 1st and 2nd, the engine rpm (in the newly selected gear) may fall below that of maximum torque if, perhaps, there is a significant grade being negotiated during a down change.

Since we know that the manufacturer has given bottom gear a ratio of 3.53 (1098cc) and 3.2 (1275cc), we can compute the expected geometrical ratio:

$$g = \frac{1}{\sqrt[3]{3.53}} = 0.657 \text{ and } g = \frac{1}{\sqrt[3]{3.2}} = 0.678$$

For a geometric progression, $s_1 = k$, $s_2 = kg$, $s_3 = kg^2$ and $s_4 = kg^3$, we thus have a comparison in gear ratios:

	1098cc			**1275cc**	
k	3.53			3.2	
g	0.657			0.679	
	Theoretical	**Actual**		**Theoretical**	**Actual**
s1	3.53	3.53		3.2	3.2
s2	2.318	2.22		2.171	1.92
s3	1.522	1.43		1.473	1.36
s4	1	1		1	1

Table 2.7.3 *Comparison of geometrical gear ratios with factory specified gear ratios.*

These data show that the gear spacing chosen by BMC are not quite in a geometric progression.

The overall gear ratio OAGR between the engine and driveshafts is computed from the gear ratios in the drive train. For example, in second gear, the gear ratios are as follows:

	Driving Teeth	Driven Teeth	Ratio
Crankshaft primary gear to 1st motion shaft	24	24	1
1st motion shaft to laygear	17	29	1.705
Laygear to 2nd speed gear	20	26	1.3
Final drive pinion to crownwheel	17	62	3.77
Overall gear ratio engine to drive shaft			8.37

Table 2.7.4 *Overall gear ratio, 2nd gear.*

In general, the road speed in km/h is calculated from the engine rpm, the rolling radius *r*, and the overall gear ratio. The rolling radius of the wheel and tyre in the present case is approximately *r* = 240mm.

$$RS = \frac{rpm(2\pi)(r)(60)}{(OAGR)1000}$$

Thus, for 2nd gear, the road speed in km/hr at 2500 rpm is found from:

$$RS = \frac{2500(2\pi)(0.240)(60)}{(8.37)1000} = 28km/hr$$

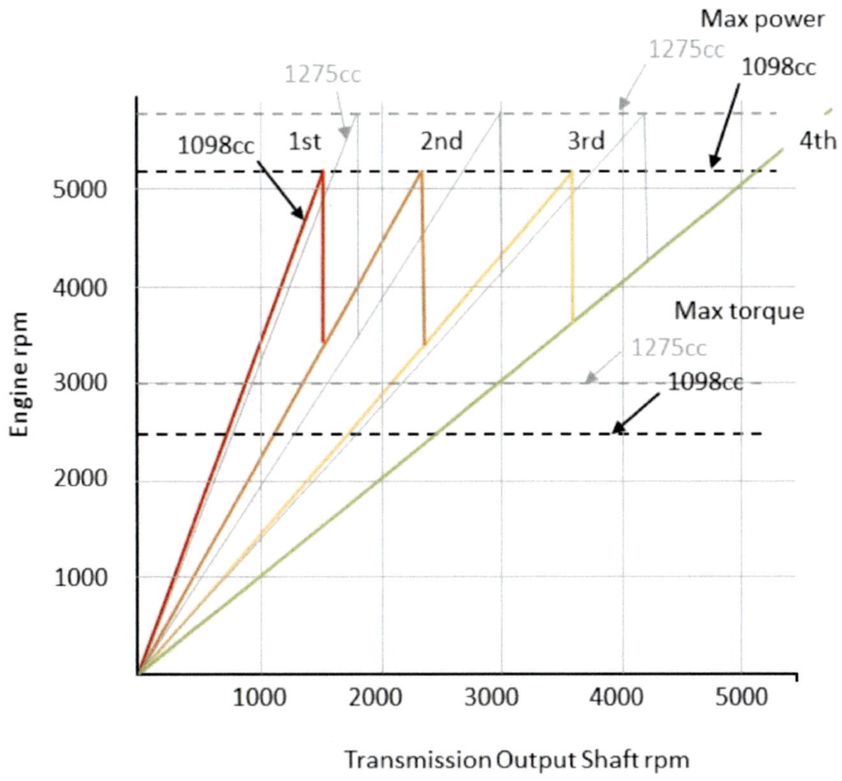

Fig. 2.7.1 *Gearbox ratio characteristics for 1098cc (heavy lines) and 1275cc (light lines).*

The figure above shows that for the 1098cc power unit, there is a noticeable gap between 2nd and 3rd gear, whereas for 1275cc, the ratios are more evenly spaced, and closer together - hence the term "close ratio".

In both cases, if gear changes are made at an engine rpm corresponding to maximum power, then the drop in engine rpm has plenty of reserve to allow for gear change times and road conditions before rpm at maximum torque occurs.

The above are of course for a 4 speed transmission. With more intermediate gears, the is even more scope for reserve torque at the number of triangular peaks increases within the rev range. In the extreme case of a continuously variable transmission, there are an infinite number of peaks in the rev range.

2.7.4 Gear Teeth

Imagine the final drive gears did not have any teeth at all, but just rubber rings around the edges – much like what is seen in a cassette tape recorder drive. It is the relative diameter of the pitch circles that gives the drive ratio and the number of teeth on each gear follows from that – with the requirement that each gear will have an integral number of teeth (no half-teeth for example) around its circumference. As well, this will in turn depend on the centre distance between the axes of the gears.

By way of an example, consider the final drive gears for a Morris Mini 850 (ADO15). The centre distance is fixed at 4.0" by the transmission case and final drive cover. How then are the whole number teeth on each gear at that centre distance for a desired final drive ratio (FDR) determined?

To see how this works, we take the centre distance as a fixed quantity – it is the distance from the centreline of the transmission 3rd motion shaft and the centre line of the crown wheel bearings. The desired final drive ratio is to be 3.765:1. This might have been chosen as a result of performance calculations shown in Section 2.8.

If the centre distance is a = 4", then the addition of the pitch circle radii must be equal to this number – drawing a line from one centre to the other.

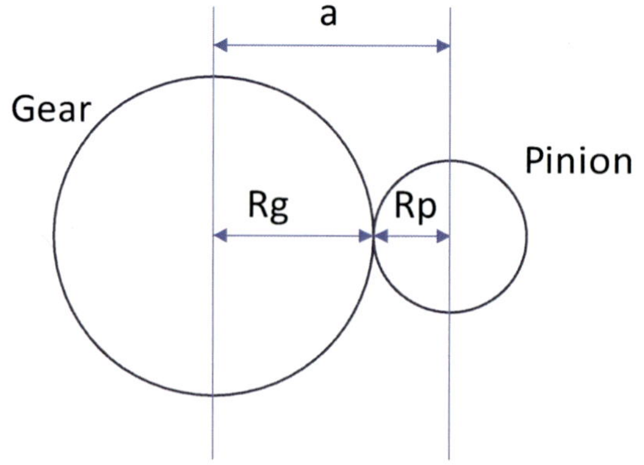

Fig. 2.7.2 *Centre distance, pinion and gear layout.*

Thus, if R_p is the pitch circle radius of the pinion, and R_g is the pitch circle radius of the gear, then:

$$R_P + R_G = a = 4"$$

Also, the final drive ratio FDR is the ratio of these pitch circle radii, and so:

$$\frac{R_G}{R_p} = FDR$$

Combining these two, we obtain:

$$a = R_P + (FDR)R_P$$
$$a = R_P(1 + FDR)$$

This is the relationship between the centre distance, the final drive ratio, and the radius (or the diameter) of the pitch circle of the pinion.

We next need to determine what whole number of teeth will fit around these pitch circle diameters to give the desired FDR. This is not so easy to do.

Gear teeth are characterised by what is known as the diametral pitch *P*. By definition, *P* is the ratio of the number of teeth *N* and the pitch circle diameter *D*. For the pinion, we have:

$$P = \frac{N_p}{D_P}$$

For what is called "conjugate action", (that is, the teeth mesh cleanly without slipping), the diametral pitch of the pinion has to be equal to the diametral pitch of the gear.

And so, for the gear,

$$P = \frac{N_G}{D_G} = \frac{N_P}{D_p}$$

The diametral pitch may also be expressed in terms of what is called the "module". The module is the inverse of the diametral pitch and is expressed in mm and is used for "metric" gears, while diametral pitch has units of inverse inches (in^{-1}) and is used for "imperial" gears.

The diametral pitch is a number that characterises the size or spacing of the gear teeth relative to the gear's diameter. The actual pitch, *p*, the distance of the arc between two similar point on the gear tooth, is:

$$p = \frac{\pi D}{N} = \frac{\pi}{P}$$

So, what we do is match the diametral pitch with the centre distance until we get a whole number of teeth.

Start with the pinion.

The pinion radius and diameter can be calculated from the centre distance *a* and the desired final drive ratio FDR using the equation above:

$$
\begin{aligned}
a &= R_P(1 + FDR) \\
R_p &= \frac{a}{1 + FDR} \\
&= \frac{4}{1 + 3.764} \\
&= 0.851" \\
D_P &= 1.702"
\end{aligned}
$$

For tooling purposes, the diametral pitch *P* for BMC gears is 10 (in inverse inch units). And so the number of teeth that would result is:

$$
\begin{aligned}
N &= P(2R_p) \\
&= 10(2)(0.862) \\
&= 17.021
\end{aligned}
$$

Now, it is obvious we cannot have 17.021 teeth on the pinion, so round this to the nearest whole number and we settle on 17 teeth for the pinion.

Using 17 teeth, and staying with 10 as the diametral pitch, the new pitch circle diameter for the pinion becomes:

$$
\begin{aligned}
D_p &= \frac{17}{10} \\
&= 1.700"
\end{aligned}
$$

What now for the gear?

Working backwards from the centre distance, the desired pitch circle radius for the gear is thus:

$$
\begin{aligned}
R_g &= a - \frac{D_p}{2} \\
&= 4 - \frac{1.7}{2} \\
&= 3.15" \\
D_g &= 6.3"
\end{aligned}
$$

Using the same value of diametral pitch *P* = 10 as for the pinion, we get the number of teeth for the gear as:

$$
\begin{aligned}
N_g &= 10(6.3) \\
&= 63
\end{aligned}
$$

This would give a final drive ratio of:

$$
\begin{aligned}
FDR &= \frac{6.3}{1.7} \\
&= 3.705
\end{aligned}
$$

There is a problem here. The actual number of teeth for this final drive gear is 64, and the actual final drive ratio is supposed to be 3.765.

Now, the above calculations were performed on the basis that everything was perfect, with no clearances between any of the gear teeth. In practice, a gear design engineer allows for a certain amount of backlash to account for thermal expansion of the parts, and space for the oil, and other manufacturing tolerances. The effect of machining the gear teeth to create these clearances effectively widens the centre distance – as if the gears were placed a little bit further apart than the ideal geometrical centres. In our case, for BMC, the effective centre distance is increased by 0.050" which makes the effective centre distance equal to 4.050".

If these calculations are repeated with a centre distance of 4.050" the number of teeth on the gear comes to exactly 64 teeth, giving a final drive ratio of 3.765 as expected.

While the above applies to spur gears, it applies as well to helical gears when we consider the geometry in the transverse plane (looking face on to the gear). We need only concern ourselves with the helical nature of the gears when it comes to designing the pressure angle, line velocity, gear loading in transverse and normal directions, and so on.

Shown below is the data taken from engineering drawings for the pinion and gear shown above.

	Pinion	Gear	
PART Nº	22A399	22A 402	
NUMBER OF TEETH.	17	64	N
* TRANSVERSE PITCH.	10	10"	P_t
PITCH CIRCLE DIAMETER.	1·7"	6·300"	D
* HELIX ANGLE.	30° L.H.	30° R.H.	φ
* NORMAL DIAMETRAL PITCH.	11·547	11·547	P_n
ADDENDUM.	·0866"	·0866"	
* NORMAL PRESSURE ANGLE	17° 29' 43"	17°-29'-43"	
* TRANSVERSE PRESSURE A.	20°	20°	
OUTSIDE DIAMETER.	1·868" / 1·873"	6.47"	
		CENTRES	

Items marked * refer to the cutter

4" centres

Fig. 2.7.3 *Final drive pinion and gear data, ADO15.*

Much the same calculations are done for all the other gear combinations in the transmission with the additional detail of pressure angle. It is interesting to note that for helical gears, the line of contact between two teeth is spread more evenly, and also diagonally, over two or three gear teeth compared to spur gears and contrary to popular opinion, are stronger than equivalent spur gears. Spur gears, or straight-cut gears, so desired by motor racing enthusiasts are merely more efficient (at the cost of increase noise) since frictional losses in the axial direction are much less compared to helical gears.

2.8 Vehicle Performance

2.8.1 Introduction

The overall performance of the vehicle on the road under different driving conditions can be shown on a vehicle performance graph. Such a graph shows road speed on the *x* axis, and both engine rpm and tractive effort on the *y* axis. We shall look at such a curve with respect to a 1098cc power unit as fitted to a Morris Mini 1100 vehicle 2/YDO5.

2.8.2 Engine Power Curve

The net power curve for a 1098cc engine is given by BMC as:

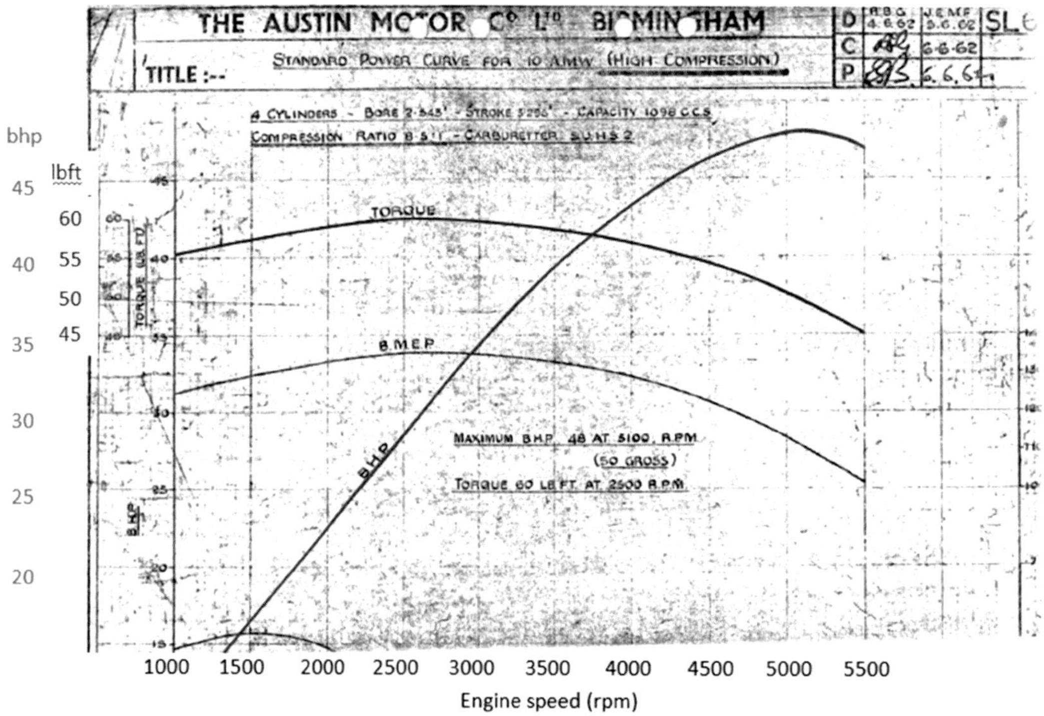

Fig. 2.8.1 *Power and torque (gross) curves for 1098cc A series engine.*

Maximum torque is shown as 60 lbft (corresponding to 81.3 Nm) at 2500 rpm and maximum power as 50 bhp (37 kW) at 5100 rpm. These are net figures. When gross figures are used, allowance must be made for various losses as shown in Table 2.8.1.

Item	Power Loss
Air cleaner	2%
Alternator	1%
Fan	2%
Distributor	3%
Exhaust	5%
Total	13%

Table 2.8.1 *Losses due to ancillaries (generalised average figures).*

Thus, the power developed at the wheels can be of the order of 20% lower than the gross engine power output.

In addition to the above, losses occur within the transmission. Here, we shall assume transmission losses of 5% in top gear and 10% in lower gears.

2.8.3 Tractive Effort

It is tractive effort that propels the car forward. Tractive effort is the force generated at the driving wheels axles whose tyres are in contact with the road. In Fig. 2.8.2, it is shown that the engine, through the transmission and final drive, applies a moment M to the driving wheels. For simplicity, we will just consider one wheel for the moment. As the moment M is applied around the stub axle, there is immediately generated a reverse reaction moment consisting of force TE acting on the tyre at the contact patch, through the rolling radius. Assuming there is no slip, and contact patch is momentarily fixed in relation to the road, the stub axle (along with the vehicle) is thus propelled forward, pivoting on the contact patch.

For the vehicle to move forward, the tractive effort must equal or exceed opposing forces arising from rolling, air and grade resistances. If the applied tractive effort exceeds air and rolling resistances, the vehicle accelerates.

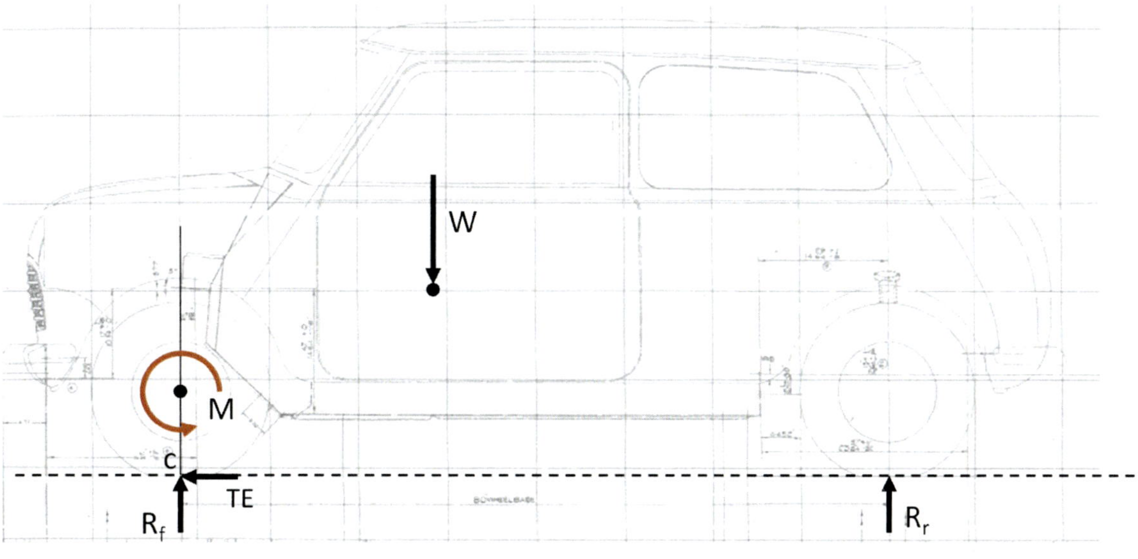

Fig. 2.8.2 *Tractive effort.*

As with all BMC front wheel drive vehicles with manual transmission, there are drop gears from the primary gear to the 1st motion shaft. The idler gear makes no contribution to the gear ratio through the drop gear train, and, in a Morris Mini, since there are 24 teeth on the crankshaft primary gear, and 24 teeth on the transmission input gear, the 1st motion shaft is therefore turning at engine speed.

Let us examine the situation in say 2nd gear and determine the tractive effort at an engine speed corresponding to that of maximum torque (T_{max} @2500 rpm) assuming a gear train efficiency η of say 90% in intermediate gears (accounting for losses).

The net engine torque at 2500 rpm is given as 60 lbft which converts to 81.3 Nm. The tractive effort TE in Newtons is thus computed from the gearbox output torque (net engine torque *T* times the overall gear ratio) divided by the rolling radius:

$$TE = \frac{(T)(OAGR)(\eta)}{r}$$

$$TE = \frac{(81.3)(8.37)0.9}{0.240} = 2.5kN$$

This is the force (shared) applied to the front tyre contact patches, the reverse reaction resulting in forward movement of the vehicle.

Values of tractive effort for a range of engine speeds and gears are shown in the Table below for final drive ratio 3.77 and wheel and tyre rolling radius 240 mm.

Gear	Gear Ratio	OAGR
1	3.53	13.31
2	2.22	8.369
3	1.43	5.391
4	1	3.770

Table 2.8.2 *Overall gear ratios all gears.*

Engine rpm	Net Torque		Road Speed	TE 1st gear	Road Speed	TE 2nd gear	Road Speed	TE 3rd gear	Road Speed	TE 4th gear
	lbft	Nm	km/h	kN	km/h	kN	km/h	kN	km/h	kN
1000	55	74.6	7.0	3.62	11.1	2.27	17.3	1.46	24.7	1.08
1500	57	77.3	10.5	3.75	16.7	2.36	25.9	1.52	37.0	1.12
2000	59	80.0	14.0	3.88	22.3	2.44	34.5	1.57	49.4	1.16
2500	60	81.3	17.5	3.94	27.8	2.48	43.2	1.60	61.7	1.18
3000	59	80.0	21.0	3.88	33.4	2.44	51.8	1.57	74.1	1.16
3500	58	78.6	24.5	3.81	38.9	2.40	60.5	1.54	86.4	1.14
4000	56.5	76.6	28.0	3.71	44.5	2.34	69.1	1.50	98.8	1.11
4500	54	73.2	31.5	3.55	50.1	2.23	77.7	1.44	111.1	1.06
5000	49	66.4	35.0	3.22	55.6	2.03	86.4	1.31	123.5	0.96
5500	45	61.0	38.5	2.96	61.2	1.86	95.0	1.20	135.8	0.88
5100	50	67.8	35.7	3.29	56.7	2.07	88.1	1.33	126.0	0.98

Table 2.8.3 *Tractive Effort (kN) in each gear for a range of engine rpm* 2/YDO5 1098cc.

2.8.4 Air Resistance and Rolling Resistance

Air resistance R_A varies with the square of the vehicle speed, the frontal area of the vehicle, the air density ρ, and drag coefficient C_d that is measured experimentally. The drag coefficient takes into account inherent influences on air resistance due to the overall shape of the vehicle (e.g. streamlining). C_d has no units, a value of 1 indicating a flat plate perpendicular to the air stream.[5]

For a Morris Mini 1100, vehicle width is approximately 1410 mm and height 1350 mm. The actual area presented to the air stream is approximately 87% of that computed from these figures to account for ground clearance and tapering of the upper body[6]. The effective frontal area is thus about 1.52 m^2. Vehicle kerb mass is specified as 633 kg to which should be added the mass of the occupants, say, 160 kg. We shall assume a value for $C_d = 0.45$ – a fairly high value given the shape of this design. Air density can be taken as 1.29 kg/m^3.

The air resistance is calculated from:

$$R_A = \frac{1}{2} C_d \rho A v^2$$

In this formula, v is in km/hr and ρ in g/L (or kg/m^3). The result is in N.

Rolling resistance R_R depends upon vehicle weight and velocity v by an empirical formula:

$$\begin{aligned} R_R &= mgC_R \\ &= mg(a + bv^n) \end{aligned}$$

In this formula, the dimensionless drag coefficient C_R comprises the constants a, b and n whose values are obtained by experiment. m is the mass of the vehicle, and g the acceleration due to gravity.

Although rolling resistance remains fairly constant for vehicle velocities up to about 80 km/h, it rises at an ever increasing rate beyond that. For wheels fitted with pneumatic tyres rolling on a hard surface, the coefficient of rolling resistance is shown to be independent of the rolling radius.

The factor a represents the road condition and is independent of velocity but does depend on tyre construction. Representative values are shown in Table 2.8.3 for cross-ply tyres. For radial tyres, the factor should be reduced by 20%.

Surface	a
Railroad	0.0045
Medium asphalt	0.0100
Ordinary macadam	0.0250
Well rolled gravel	0.0254
Hard, dry clay	0.0450
Medium cobbles	0.1600
Loose sand	0.2500

Table 2.8.3 *Factor a for coefficient of rolling resistance for various road surfaces (cross-ply tyres).*

The dynamic factor b depends on the construction of the tyre, and in particular, it's speed rating. An SR tyre has a much higher value compared to a HR tyre. For a standard cross ply tyre, a value of $b = 11.7$ is reasonable.

[5] More like 1.25 due to turbulence around the edges which results in a wake of greater area than that presented.

[6] A factor of 80% is usually applied but for the boxy shape and low ground clearance of this vehicle, a figure of 87% has been used.

Tyre Type	Speed Rating	Max Speed (km/h)	*b*
Cross-ply	-	150	11.7
Cross-ply	L	160	8.2
Cross-ply	S	180	4.2
Cross-ply	H	210	2
Radial	-	170	4.5
Radial	SR	180	3.3
Radial	HR	210	1.4
Radial	VR	240	0.7

Table 2.8.4 *Factor b for coefficient of rolling resistance.*

The index *n* is determined from the equation above fitted to experimental data and influence how steeply the rolling resistance rises with increasing velocity. The overall rolling resistance at higher speeds is particularly sensitive to the value of *n*. A value of 5.6 is reasonable.

m is in kg, and *v* is in km/hr. *g* is the acceleration due to gravity, taken to be 9.8 m/s^2. Values calculated for a range of road speeds are in Table 2.8.5 for m = 793 kg, frontal area 1.41 m^2, C_d = 0.45, a = 0.018 (cross ply[7]), b = 11.7, n = 5.6. The mass used here is the kerb mass 633 kg plus the mass of two occupants estimated at 160 kg.

Road Speed	Rolling Resistance Coefficient	Rolling Resistance	Air Resistance	Total Road Resistance
km/h	Cr	kN	kN	kN
0	0.0180	0.140	0.000	0.140
12.0	0.0180	0.140	0.005	0.145
24.0	0.0180	0.140	0.020	0.160
36.0	0.0180	0.140	0.044	0.184
48.0	0.0180	0.140	0.079	0.219
60.0	0.0181	0.141	0.123	0.264
72.0	0.0184	0.143	0.177	0.320
84.0	0.0189	0.147	0.241	0.387
96.0	0.0199	0.155	0.314	0.469
108.0	0.0217	0.168	0.398	0.566
120.0	0.0246	0.191	0.491	0.682
130.0	0.0283	0.220	0.576	0.797

Table 2.8.5 *Air and rolling resistance, 0% grade.*

2.8.5 Grade Resistance

When travelling up or down a grade, a proportion of the vehicle mass contributes to the resistance or addition to tractive effort depending on the angle of the grade. This proportion is found from simple geometry, such that the grade resistance F_G is:

[7] a = 0.014 for radial ply tyres

$$F_G = mgsin\theta$$

Values calculated for a range of grades are in Table 2.8.6 for *m* = 793 kg.

Grade %	theta (deg)	GR (kN)
5	2.86	0.388
15	5.71	1.152
25	8.53	1.885
35	11.31	2.567

Table 2.8.6 Grade resistance for *m* = 793 kg for a range of grades.

The grade is the angle at which the road makes with the horizontal. More commonly, the grade is expressed as a ratio, for example, 1 in 3. This is the tan of the angle in the sense that the road rises a height of 1m over a horizontal distance of 3m. Alternately, the gradient may be expressed as a percentage, for example, 1:3 = 33% grade. In this scheme, a 100% grade is one at 45° to the horizontal.

2.8.6 Performance Curves

All the above data may be then plotted to give an overall view of vehicle performance. Putting all these data together, we obtain a good picture of the characteristic performance of the vehicle under a wide range of operating conditions.

Thus, for a Morris Mini 1100, 2/YDO5, Fig. 2.8.3 shows tractive effort plotted against road speed. The tractive effort plotted in Fig. 2.8.3 is the net tractive effort available to the wheels taking into account transmission losses on level ground. On the same graph, the total road resistances are plotted for a range of grades. These take into account rolling and air resistances.

As can be seen from this figure, one cannot expect to maintain a road speed of say 115 km/h in top gear going up a gradient of more than about 5% since there is not enough tractive effort from the engine to overcome the combined rolling, air and gradient resistances.

On level ground with no headwind, it would seem that maximum speed estimated from Fig. 2.8.3 is about 130 km/h (80 mph). Road reports of the time show a maximum speed on a test vehicle of 77.6 - 80 mph[8].

[8] 77.6 mph Modern Motor, 80 mph Wheels Jun 1969

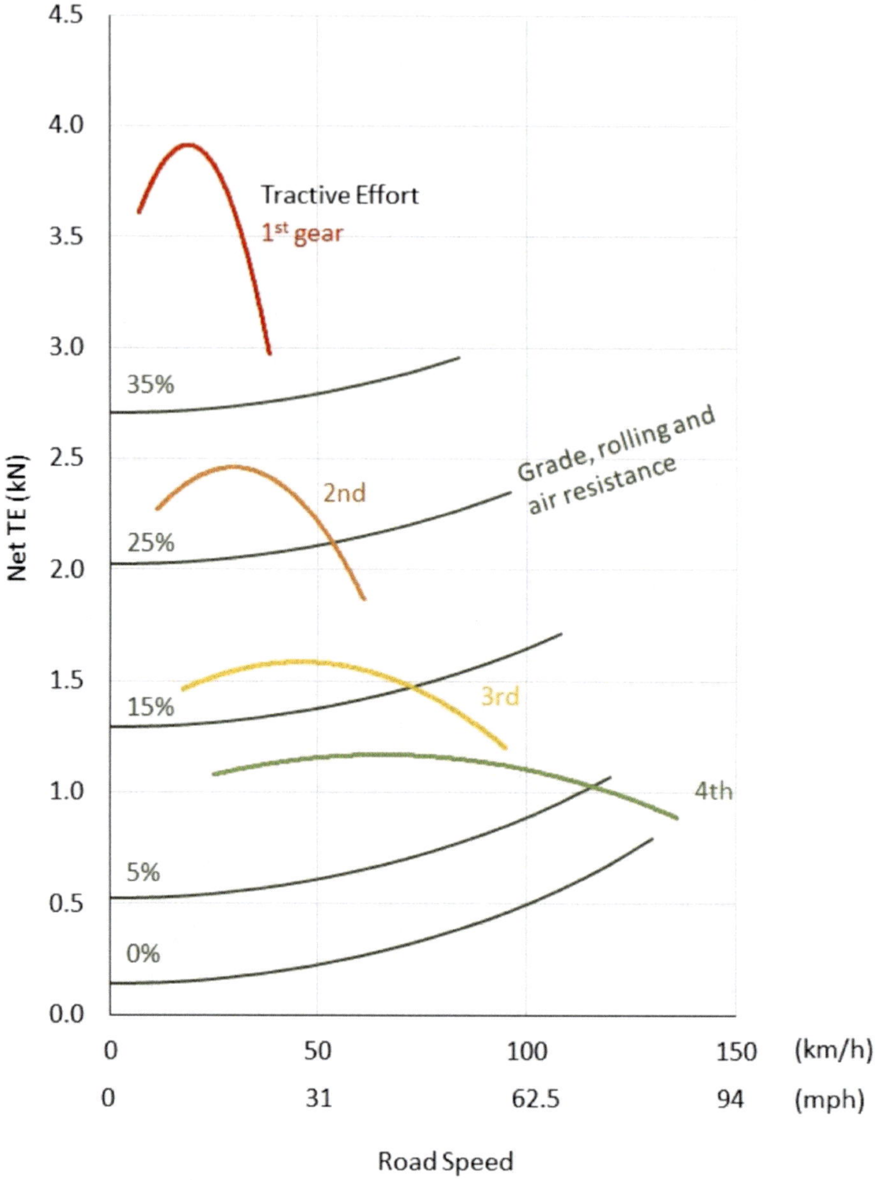

Fig. 2.8.3 *Vehicle Performance Graph Tractive Effort vs Road Speed, 1098cc 2/YDO5.*

The resistance to motion increases as the vehicle mass increases, and particularly so for gradient resistance. We might wish to know the maximum climbing gradient. This occurs in first gear at maximum torque condition, at the gross vehicle mass. Calculations performed at a gross vehicle mass of 950 kg show that the maximum climbing gradient is of the order of 39%. However, the maximum climbing gradient is somewhat more optimistic that the maximum starting gradient. It is generally accepted that the torque to start on a gradient is a factor of x1.3 that required for steady state motion. In the present case, the maximum starting gradient is therefore about 25%.

2.8.7 Vehicle Acceleration[9]

So far, the vehicle performance characteristics discussed have been steady state values – that is, what the vehicle maximum speed might be, gradient capability, and so on. It is of course desirable that a vehicle have a reasonable top speed, but not take all day to get there. Thus, it is important to study the dynamic effects of acceleration.

The tractive effort shown in Fig. 2.8.3 is the tractive effort available to accelerate the vehicle. The acceleration achieved depends on the tractive effort available after overcoming road resistance, and the mass of the vehicle. However, as far as acceleration is concerned, it is necessary to consider the vehicle mass *and* rotational moments of inertia of components like the engine flywheel and road wheels. All these effects can be combined into what is called an "effective mass".

2.8.7.1 Effective Mass

When a motor vehicle is accelerated, not only does the vehicle mass increase in velocity as a whole, but rotating parts within it also increase in rotational speed, or more precisely, angular velocity. The engine generally revolves faster, the wheels rotate faster, etc. All of these increases in angular velocity require energy and in order to compute what is left over for linear motion of the vehicle, must be accounted for.

It is convenient in working out the effect of rotating masses to have as a reference, the road wheel. Let us imagine that the engine and associated drive train components are frictionless, and for the moment, all the wheels are considered as one single wheel, which is connected via a gear train to the crankshaft.

Since our point of view is the wheel, we apply a force to the road wheel at its circumference and compute how much torque would be required to produce a certain angular acceleration of the crankshaft.

As with all rotational motion, it is the moment of inertia I and angular acceleration α that take the places of mass and acceleration in linear motion. In general, when an unbalanced torque T is applied to a body with moment of inertia I, the resulting angular acceleration α can be found from:

$$T = I\alpha$$

For now, we focus on the angular acceleration of the engine components. We will later add in the torque to be applied to the road wheels to overcome the rotational inertia of the wheels and gear train.

There is usually a gear reduction from the crankshaft to the wheels, and so engine torque is multiplied upwards by the transmission and final drive ratios by the time it is applied to the wheels. Of course the wheels turn at a correspondingly lower rotation speed by the same factor. Conversely, any torque T_E applied to the wheels will result in a lower value of torque applied to

---

[9] Acknowledgement: Ray Habgood

the crankshaft. Therefore, if a torque T_E is applied at the wheels, then a lower torque T_C is applied to the crankshaft.

$$T_C = \frac{1}{R_E} T_E$$

where R_E is the engine speed ratio:

$$R_E = \frac{Engine\ Speed}{Wheel\ Speed}$$

Or, from the point of view of the wheels,

$$T_E = R_E T_C$$

R_e is the overall gear ratio from engine to wheels, and thus includes the selected gear ratio (1st, 2nd, etc) and the final drive ratio.

Now, when an unbalanced torque T_E is applied to the wheels, and thence a torque T_C to the crankshaft, there results an angular acceleration α_C of the crankshaft. Since there is a gear reduction between the crankshaft and wheels, the angular acceleration of the crankshaft α_C is more than that of the wheels α_W such that:

$$\alpha_C = R_E \alpha_W$$

The relationship between torque and moment of inertia given above means that from the point of view of the wheels, the torque T_E required to accelerate the engine components is:

$$\begin{aligned} T_E &= R_E I_C \alpha_C \\ &= R_E I_C R_E \alpha_W \\ &= I_E \alpha_W R_E^2 \end{aligned}$$

By a similar argument, the torque required to accelerate the geartrain and associated components is:

$$T_G = I_G \alpha_W R_G^2$$

Where here, I_G is the moment of inertia of the gearbox rotating parts, and R_G is the gear ratio of the final drive:

$$R_G = \frac{Gear\ Speed}{Wheel\ Speed}$$

In the gearbox, various members are rotating at different speeds, but can be considered to include all the gears that rotate with the final drive pinion inside the gearbox as distinct from those rotating an engine speed (whose moment of inertia is included in T_E). The value of R_G is thus the final drive ratio.

For the rotation of the wheels, and connected components like the drive shafts and final drive crown wheel, there is no gear reduction involved since all these rotate together at the same angular velocity and so the torque required to accelerate these components is simply:

$$T_W = I_W \alpha_W$$

Now that the various torques at the wheels required to serve the rotating components have been described, it remains for us to consider the torque required to produce a linear acceleration of the vehicle mass, T_L.

The linear acceleration a of the vehicle mass m is expressed:

$$F = ma$$

Where F is the force acting on the wheel hub, and whose reaction force is at the contact between the tyre and the road.

Expressed in terms of a torque acting through a radius R_R (the rolling radius of the wheel and tyre), the torque T_L is thus:

$$\begin{aligned} T_L &= FR_R \\ &= maR_R \end{aligned}$$

We can look at this in terms of the angular acceleration of the wheel, but bearing in mind we are doing so from the point of view of the linear acceleration of the vehicle mass m, and not the rotating mass of the wheel and associated components which has been described previously.

The linear a of the mass m and angular acceleration of the wheel α_W are related by:

$$a = R_R \alpha_W$$

And so, in terms of an unbalanced torque, T_L, acting around the wheel hub, we have:

$$\begin{aligned} T_L &= mR_R \alpha W R_R \\ &= \alpha_W m R_R^2 \end{aligned}$$

Thus, to accelerate the vehicle an amount a, we require the following torque total T to be applied at the wheel.

$$\begin{aligned} T &= T_E + T_G + T_W + T_L \\ &= I_E \alpha_W R_E^2 + I_G \alpha_W R_G^2 + I_W \alpha_W + \alpha_W m R_R^2 \\ &= \alpha_W (I_E R_E^2 + I_G R_G^2 + I_W + mR_R^2) \end{aligned}$$

The linear acceleration a can be written:

$$\begin{aligned} a &= R_R \alpha_W \\ &= R_R \frac{T}{(I_E R_E^2 + I_G R_G^2 + I_W + mR_R^2)} \end{aligned}$$

Or,

$$\frac{(I_E R_E^2 + I_G R_G^2 + I_W + mR_R^2)}{R_R} a = T$$

In terms of what is essentially the tractive effort, F, and the torque T, then $F = R_R T$ and so

$$F = \left[\frac{(I_E R_E^2 + I_G R_G^2 + I_W + mR_R^2)}{R_R^2}\right] a$$

The square-bracketed term is called the effective mass m_{eff}, and represents the total mass, in a linear sense, that has to be accelerated by the tractive effort F. That is, the effective mass is the mass of the vehicle plus the additional mass which takes into account the acceleration required to increase the angular velocity of the rotating parts. Since the effective mass contains a factor which represents the gear and final drive ratios, then it is different for each gear[10]. The equation shows that the effective mass m_{eff} it greater than the vehicle mass m.

2.8.7.2 Acceleration Times

As an example of the above, the following data is applicable to a Morris Mini 1100 vehicle, 2/YDO5.

Quantity	Symbol	Value	Unit
Moment of inertia of engine and associated components	I_E	0.085	kg m^2
Moment of inertia of gearbox and associated components	I_G	0.004	kg m^2
Moment of inertia of 4 wheels, 4 tyres, 2 drive shafts and final drive components	I_W	1.334	kg m^2
Gear ratios			
1st		3.53	
2nd		2.22	
3rd		1.43	
Top		1.000	
Final drive ratio		3.77	
Vehicle kerb mass + Occupants	M	793	kg
Tyre rolling radius	R_R	0.240	m

Table 2.8.7 *Vehicle data 2/YDO5 Morris Mini 1100.*

The effective mass for determining acceleration of the vehicle for each gear ratio can be computed thus:

$$m_{1st} = \frac{(I_E R_E^2 + I_G R_G^2 + I_W + mR_R^2)}{R_R^2}$$

$$= \frac{(0.085)\left((3.53)(3.77)\right)^2 + (0.004)(3.77^2) + (1.334) + (793)(0.240^2)}{0.240^2}$$

$$= 1139.2kg$$

And by similar computations for other gears:

$$m_{2nd} = 963.7kg$$

$$m_{3rd} = 869.5kg$$

$$m_{Top} = 872.2kg$$

[10] It can be seen that the addition of free-wheeling front hubs on a four wheel drive vehicle would offer substantial reduction in effective mass since rotation of the front drive shafts and final drive need not be considered when the hubs are disconnected from the road wheels.

Can we now compute acceleration times? To answer this question, let's assume that gear changes take place at 5100 engine rpm (maximum power).[11] Fig. 2.7.1 shows the relationship between road speed and engine speed. In first gear, we can expect to achieve a road speed of 35.4 km/hr at 5100 rpm based upon the gear ratios involved. The maximum possible speeds in all gears at an engine speed of 5100 rpm are shown in Table 2.8.8.

Gear	Road speed (km/h) at 5100 rpm
1	35.4
2	53.4
3	82.9
4	118.6

Table 2.8.8 *Road speed at 5100 rpm for 2/YDO5.*

The question is, how long does it take to achieve these speeds?

The effective tractive effort available for acceleration is the difference between that provided by the engine, less that required to overcome air, rolling and grade resistances. For the example to be shown here, we will assume 0% grade, level ground. The data shown in Fig. 2.8.3 can be replotted by subtracting off the air and rolling resistance terms so that the graph provides the excess, or net, tractive effort available for acceleration. The revised graph, for zero road gradient, is shown in Fig. 2.8.5. where the speed range gas extended to an engine rpm of 5500 as an absolute maximum.

As can be seen in Fig. 2.8.5, the net tractive effort varies considerably over the speed range due to the variation in engine torque and combined resistances to motion. These variations vary quite smoothly and so can be modelled mathematically by curve fitting. In the case of the net tractive efforts shown in Fig. 2.8.5, least squares curve fitting yields the following expressions where *v* is in m/s.:

Gear	Net TE
1	$TE = -33.75v^2 + 366.06v + 2695.7$
2	$TE = -8.775v^2 + 132.99v + 1652.6$
3	$TE = -2.736v^2 + 55.73v + 1021$
4	$TE = -1.3947v^2 + 31.484v + 704.38$

Table 2.8.9 *Fitting equations for tractive effort for data shown in Fig. 2.8.5.*

Since Fig. 2.8.5 and the equations shown in Table 2.8.9 express tractive effort TE_{net} *change* as a function of velocity, the integral of these equations is of physical significance, it has units of power. The integral

$$P = \int_{v_1}^{v_2} TE(v)dv$$

[11] We don't have to change gears at max power rpm, since the engine is usually capable of going beyond this to some safe maximum. But, because of the fall-off in torque as rpm rises, although the vehicle may keep travelling at a faster road speed, there is less available tractive effort for overcoming road resistance and any desired acceleration. We eventually get to the point where the power developed is balanced by the power consumed by road resistance and maximum speed is reached.

represents the area under the plot of net TE_{net} against v, and when this area is divided by the velocity range $v_2 - v_1$, gives the effective constant tractive effort TE_{av} applied during the time taken for the vehicle to change velocity in that velocity increment[12]. Thus, for say for first gear, and working in velocity units of m/s, we have for 0 to 9.84 m/s (0 – 22 mph, 0 – 35.4 km/h):

$$TE_{eff} = \frac{\int_0^{9.84} -33.75v^2 + 366.06v + 2695.7\,dv}{v_2 - v_1}$$
$$= \left[-\frac{67.4}{3}v^3 + \frac{327.1}{2}v^2 + 2695.7v\right]_0^{9,84}\left(\frac{1}{9.84}\right)$$
$$= 3407.7N$$

Since we are changing gears at 5100 (maximum power) we can now perform an integration for the range 35.4km/h to 53.4km/h for second gear. Continuing on for 3rd and 4th (top) gear, we obtain the results shown in Table 2.8.10. The integration has to be taken for only positive values of TE_{net}. When TE_{net} falls below zero, there is insufficient tractive effort to produce motion of the vehicle.[13]

Quantity	1st Gear	2nd Gear	3rd Gear	4Th Gear	Units
TE_{eff}	3407.7	1939.8	1080.2	481.7	N
M_{eff}	1062.5	913.4	856.3	835.6	kg
Velocity	35.4	53.4	82.9	118.6	Km/h
Acceleration	3.21	2.12	1.26	0.58	m/s²
Increment Time	3.07	2.35	6.50	17.20	sec
Accumulated time with GC	3.07	6.02	13.12	30.92	sec
Accumulated time without gear change delay	3.07	5.42	11.92	29.12	sec
Increment distance	15.1	29.0	123.0	481.4	m
Accumulated distance travelled	15.1	44.1	167.1	648.5	m
Power	12.50	16.09	17.64	14.78	kW

Table 2.8.10 *Results of performance calculations through the gears at wide open throttle with gear changes at maximum power engine rpm.*

So, for a particular speed range in a particular gear, we can calculate the acceleration from the net value of TE_{eff} as applied to the effective mass m_{eff}, and from there, by ordinary kinematical equations, calculate desired quantities of interest.

For example, the acceleration during a period in each gear is found from:

[12] Via the mean value theorem of calculus.

[13] There is a further complication when determining the effective tractive effort at a specified distance, e.g. a standing 400m. With the method described above, the value of TE_{eff} would be calculated from the integral of the lowest speed for that gear to the highest (or maximum speed in the case of the vehicle reaching its maximum speed). What is needed is the integral from the lowest speed for that gear to the terminal speed – which is not known in advance. The procedure then is to calculate an approximate terminal speed based on these wider limits of integration, and then recalculate that speed (and travel time) with that approximate terminal speed. This captures more closely the shape of the TE_{net} curve through that last increment of travel.

$$a = \frac{TE_{eff}}{m_{eff}}$$

The time taken is:

$$t = \frac{v_2 - v_1}{a}$$

And the distance travelled:

$$d = v_1 t + 0.5at^2$$

Knowing the accumulated time to reach, say, 118.6 km/h, then the kinetic energy expended over that time indicates the power available used in accelerating the vehicle.[14]

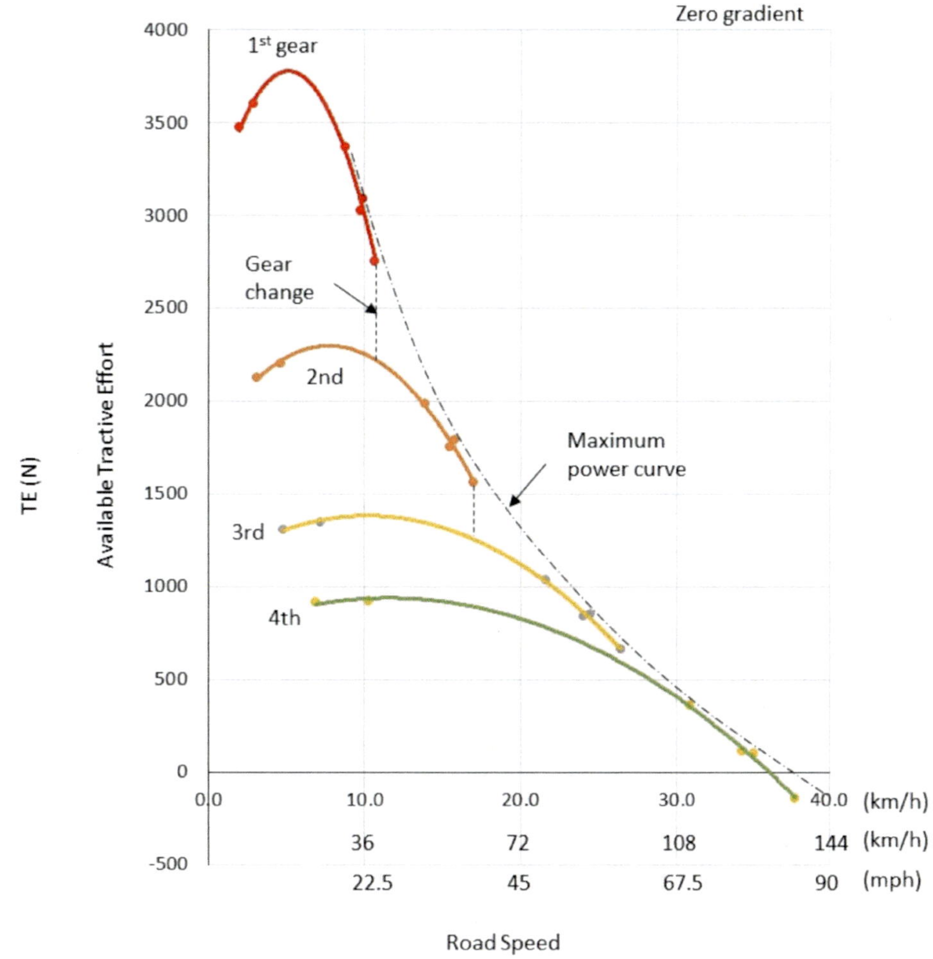

Fig. 2.8.5 *Net tractive effort for data plotted in Fig. 2.8.3 for zero gradient.*

[14] The figures indicated in Table 2.8.10 for power have been calculated without the gear change time included. This is the "excess" power required to accelerate the vehicle mass calculated from the kinetic energy of the vehicle over the accumulated time.

As can be seen in Table 2.8.10, we expect the vehicle to take 3.1 seconds to reach 35.4km/h in 1st gear. When added cumulatively, the travel times for 35.4km/h, 53.4km/h, 82.9km/h and 118.6km/h through the gears can be readily predicted. An allowance of 0.6 seconds for each gear change has been made for the accumulated travel times shown in Table 2.8.10.

These data can also be expressed in terms of vehicle speed vs time and this provides a graphic measure of vehicle performance which can be used for comparing different vehicles or test conditions. In Fig. 2.8.6, calculations for the Morris Mini range of vehicles have been shown to highlight the difference in performance of the various models.

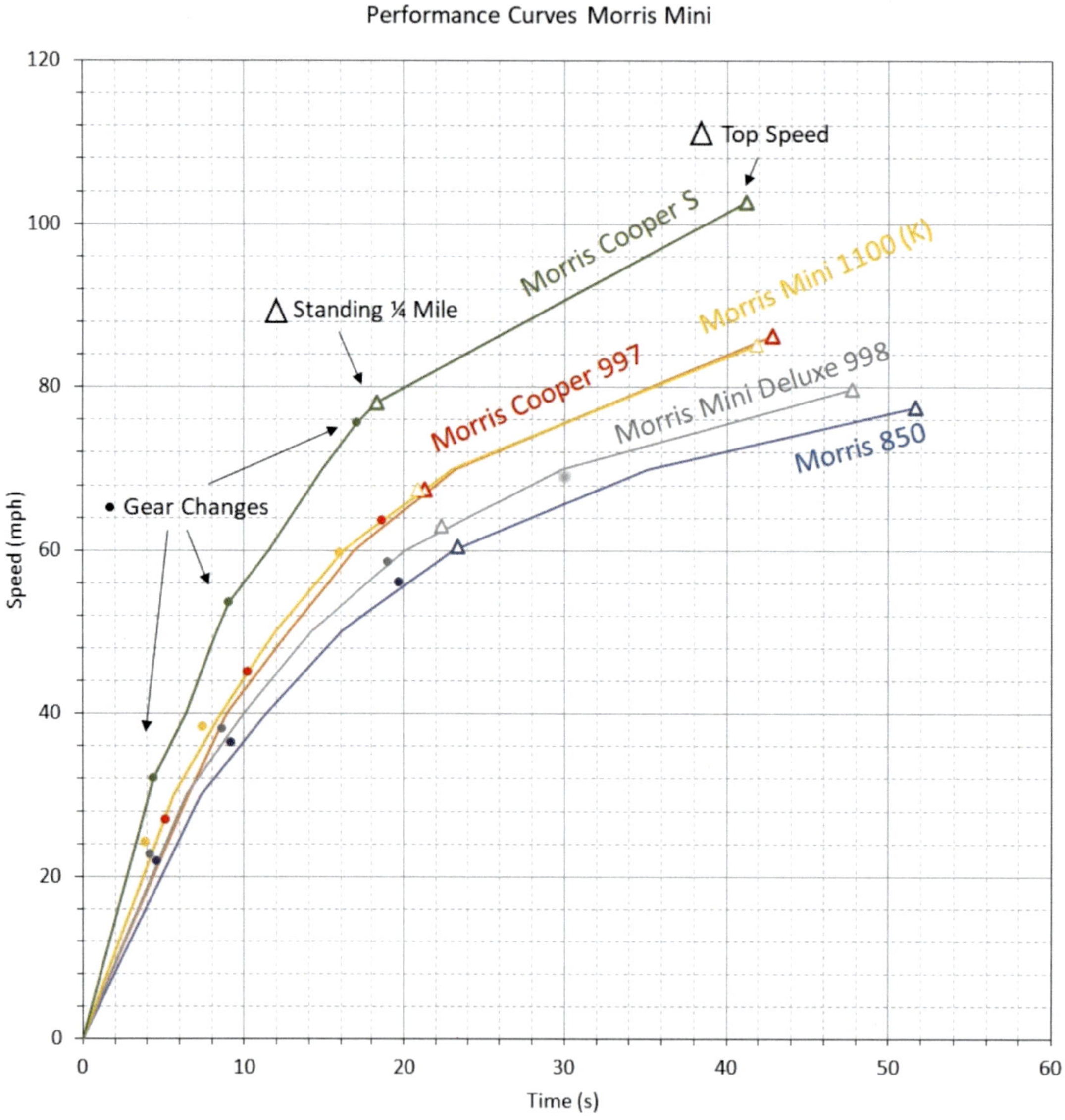

Fig. 2.8.6 *Vehicle Performance Graph Tractive Effort vs Road Speed, 1098cc 2/YDO5.*

Comparison with Modern Motor and Wheels Magazine June 1969 test data for the Morris Mini 1100 2/YDO5 is shown in Table 2.8.11.

Quantity	MM	Wheels	Calculated	Units
Top Speed	77.6	80	85.3	mph
1/4 mile	20.5	20.9	20.9	sec
0-30	4.8	4.8	5.6	sec
0-40	7.4	7.8	8.6	sec
0-50	12	12	11.9	sec
0-60	18.6	17.6	16.2	sec
0-70	30	25	23.1	sec
10YB 10YC Engine 3.77 FDR				

Table 2.8.11 *Through-the-gears acceleration, Morris Mini 1100, 2/YDO5.*

Also shown in Table 2.8.11 is a comparison with the computed data for estimated top speed, standing quarter mile and acceleration times through various speed ranges.

The correspondence between calculated and test data is reasonably good, especially when one considers that the gear change points under test are not known, the effect of any wheel spin upon take-off not accounted for, tail or head wind and gradient effects on test, reaction time when starting off, actual tyre rolling radius, and state of engine tune.

The vertical dashed lines in Fig. 2.8.5 indicate the gear change points if gear changes were made at an engine rpm corresponding to maximum power. The tangential curve shown in Fig. 2.8.5 indicates the line of maximum power that could be used, for example, the limiting case of a continuously variable transmission.

The transmission therefore allows us to have a reasonable amount of acceleration in lower gears where it is mostly needed at the expense of top speed in lower gears, while the higher gears provide less reserve of tractive effort but allow higher vehicle speeds.

These types of calculations show quantitatively those factors which influence vehicle performance. When a decision is made to manufacture a particular model, a manufacturer cannot afford to wait to build a prototype and then determine its performance for comparison with competitor products. Calculations like those above enable a manufacturer to tune design parameters before a vehicle is made to suit the market niche being targeted.

Another useful outcome of these calculations is to predict the effect of various modifications on performance. For example, by lightening the flywheel, not only is the static mass reduced, but also the effective mass, and the effect on acceleration times can thus be calculated, and as well, from Section 2.4, the change in engine vibration.

It is interesting of course that as a matter of history, Leonard Lord knew what he wanted to achieve marketwise, and set Issigonis to design something with almost a free hand. Issigonis was a master at keeping the weight down and interior dimensions up, and it wasn't until the prototype was built was the decision made to go ahead. Interestingly, the original engine capacity had to be reduced by about 100cc so that the vehicle would not be considered over-powered.

CHAPTER 3.
EMISSION CONTROL

3.1 Emission Control

3.1.1 Introduction

One of the most pressing problems of our time is the issue of exhaust emissions from motor vehicles. Since the 1960s, the number of motor vehicles has increased to the extent that they are a major source of air pollution.

Before the introduction of emission control systems, a typical motor vehicle would emit unburnt hydrocarbons (HC) of which about 50% would be found in the exhaust gases, 20% from crankcase vent, 10% from the carburetter float bowl vent, and 20% form the fuel tank vent. The presence of these hydrocarbons in the atmosphere, together with the oxides of nitrogen (NO_x) emitted from the exhaust system, are the ingredients for photochemical smog.

Although the combustion process ideally produces just carbon dioxide and water, in practice, exhaust gases contain carbon dioxide, carbon monoxide (CO), oxides of nitrogen, carbon particles, unburnt hydrocarbons, lead (when it used to be added to petrol), and of course, water.

To reduce the formation of harmful acids and dilution of engine oil, crankcases have been traditionally vented via a flow-through air system – the simplest of which was a tube pointing downwards into the air stream from the crankcase, and a vented oil filler cap. Air flow from vehicle motion passing the end of the tube creates a suction which draws a flow of fresh air through the crankcase via the filler cap. In time, this system underwent a significant improvement with the introduction of positive crankcase ventilation where instead of venting crankcase gases out to the atmosphere, they were fed back into the air cleaner and burnt in the combustion chamber. A PCV valve regulated the flow of air through the system.

While a worthy undertaking, much more sophisticated methods are required to reduce exhaust and evaporative emissions. In Australia, various state and federal bodies legislated the requirements for both type of emissions, chiefly through the NSW Clean Air Act, and the Commonwealth Australian Design Rule (ADR) system.

Initially, from 1974, the concentration of CO at idle was limited to 4.5%. In 1976, the levels of CO, HC and NO_x were constrained to limits when vehicles were tested in a prescribed manner on a dynamometer in a testing laboratory. From 1975 onwards, the lead content in petrol was required to be reduced, and in 1985, unleaded petrol was prescribed for all new vehicles. The introduction of unleaded fuel allowed vehicles to be fitted with catalytic converters in the exhaust system which would reduce the levels of CO and HC appreciably. A summary of the legislated levels of emissions in grams per kilometre for petrol driven passenger cars and derivatives is shown in Table 3.1.1.

Legislation	**g/km**	**g/km**	**g/km**
	CO	HC	NOx
1976 ADR27A	24.2	2.1	1.9
1982 ADR27B	24.2	2.1	1.9
1983 ADR27C	24.2	2.1	1.9
1986 ADR37 (92 Octane unleaded)	9.3	0.93	1.93

Table 3.1.1 *Emissions limits in grams per kilometre set by various Australian Design Rules.*

It can be seen that the focus of these legislated limits is to reduce CO and HC rather than NO_x. Such reductions reduce the levels of photochemical smog and, with the addition of a catalytic converter, allow engines to be designed to operate more efficiently.

There is a trade-off involved with the level of emissions and optimum air fuel ratio (stoichiometric) is shown in Fig. 3.1.1.

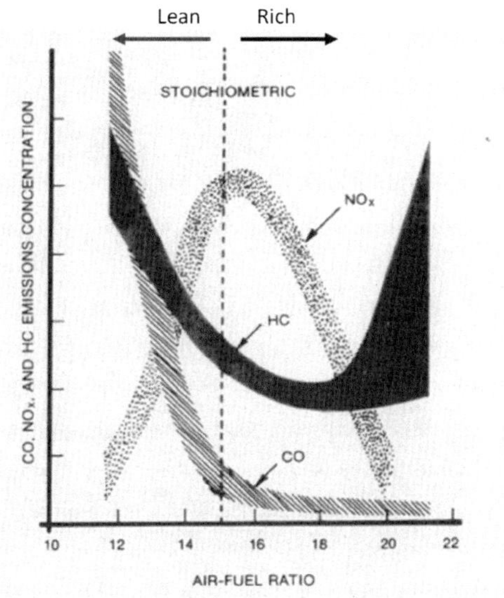

Fig. 3.1.1 *Relationship between CO, HC and NOx with air fuel ratio.*[15]

Early attempts at emissions control were directed at reducing the occurrence of predominantly CO and HS before they were formed in the combustion chamber. Mixtures became lean, carburetter jetting more precise, and as such, engines required more careful and periodic maintenance. In practice, emissions were reduced substantially if the engine was kept in tune but were often worse emitters if maintenance was neglected. As a consequence, between 1976 and prior to the introduction of catalytic converters in 1986, vehicles gained a reputation for poor driveability, over-heating, running-on, and high fuel consumption.

To overcome some of these problems, many manufacturers turned to the use of an air-injection system. An engine-driven air pump would inject air into the exhaust just behind the exhaust valve.

[15] NSW State Pollution Control Commission

At this point, unburnt hydrocarbons and carbon monoxide are still hot enough to burn, and such a system did work, at the expense of loss of mechanical efficiency associated with driving the air pump. The philosophy was a change from preventing the undesirable products of combustion forming, to cleaning them up afterwards. This later found its ultimate expression in the form of a catalytic converter which would effectively "burn" CO and HC in a muffler-shaped device in the exhaust system – close to the engine, but not right at the exhaust valve. As well, with the introduction of more electronics in motor vehicles, an oxygen sensor was added to the exhaust system whose output was fed back to the fuel injection computer (by now standard equipment) thus forming a closed-loop system.

For BMC/Leyland vehicles, emission controls followed legislated requirements beginning with the introduction of a vacuum controlled PCV system on most MKII models of the front wheel drive range. That is, from about 1969. ADR27a was introduced in July 1976 and by then, Mini and Moke production was taking place at Enfield. The 998cc EPAI (Exhaust Port Air Injection) power unit was imported CBU and fitted complete. No adjustments or alterations were permitted to be made before fitment to the vehicle. This even included removing the rocker cover which, for the Leyland Mini S models (YDO22), had been painted silver birch to distinguish it from the lower spec YDO21 model. Thus, power units were finished in black all over from that time onwards.

Leyland Mini and Moke vehicle in Australia were not fitted with an exhaust gas recirculation (EGR) valve nor a catalytic converter.

3.1.2 Crankcase Ventilation

There are four types of crankcase ventilation systems to be found in BMC vehicles. A description of each follows, but we must first remember what the ventilation system is for. As is well known, the seal of the piston rings is not perfect, and during compression and combustion strokes, there is blowby from the combustion chamber into the crankcase. It is thought that some 80% of this blowby is unburnt fuel, the remainder being products of combustion – a large part of which is water. If these gases and water vapour were to remain in the crankcase, they would form sludge with the oil which is of course undesirable, not to mention the presence of corrosive compounds. The engine ventilation system is supposed to circulate a continuous stream of fresh air through the crankcase so as to remove the blowby gases before they have time to do any damage.

The first method of crankcase ventilation (used by many cars of the early 60s) was to have a vented oil filler cap and what is called a "draft tube" which was connected to the crankcase and pointed downwards into the under-car air stream so that passage of the car would create a suction effect thus drawing fresh air though the cap, through the crankcase, and out the tube. Effective, but hardly environmentally friendly.

The next method, such as that found on BMC vehicles of the 1960s is similar in operation but the air goes the other way. The oil cap is sealed. A pipe joins the rocker cover to the air cleaner intake, and there is a tube in the engine side plate. This tube has all the appearance of being crimped off, but in fact, there is a very small gap in the end of the crimp. Air is drawn through the crimped end, through the crankcase, and out the rocker cover into the air cleaner. This also works quite well, but is unregulated – so that much the same amount of ventilation occurs at idle as compared to full throttle. In this system, removing the oil filter cap shouldn't make any different to the idle rpm. The oil cap is plastic, short neck, and non-vented.

The third type of system is one which involves the PCV diaphragm valve which can be found on a MKII Cooper S or an Austin 1800. In this system, there is a vented filler cap and a breather separator in the timing gear cover. Air travels into the rocker cover via the vented cap, through the crankcase, through the oil separator in the breather, up to the control valve, and then into the inlet manifold. The control valve is needed because (unlike the air cleaner depression), the vacuum level can be quite high. What is unusual about this valve compared to just an ordinary PCV valve is that it is controlled by crankcase pressure, or more correctly, crankcase vacuum. At idle, there is a certain amount of crankcase vacuum present due to the action of the PVC system as a whole, and the valve is designed to limit the amount of ventilation applied. A feedback loop. At higher engine loads, the blowby increases and so the vacuum falls, and the valve opens more to increase the ventilation effect to accommodate the extra blowby that needs to be handled. In this system, removing the oil filter cap should result in a slight increase in idle rpm. It should not stall. The filer cap has a long neck, and is vented – either through two holes in the top surface, or slots under the rim. Inside, there is a wire mesh filter.

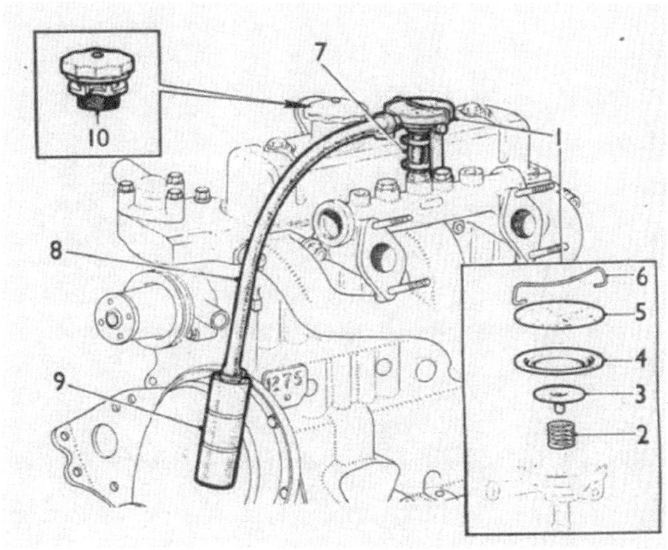

Fig. 3.1.2 *PCV System Arrangement.*

The fourth type worthy of mentioning involves a connection to the carburettor body. The ventilated air is fed into a white plastic pipe connector which is positioned upstream of the throttle disc and downstream of the carburetter piston. Since the velocity of the air flow through the carburettor past this port depends on the throttle opening, then resulting depression (similar in action to the original draft tube) serves to control the amount of crankcase ventilation depending on engine load[16]. Fresh air enters via the rocker cover.

[16] Note, the velocity of the air flow at the crankcase ventilation port is not a constant, unlike the velocity of the air past the main jet – which by virtue of the piston, *is* a constant.

3.1.3 Intake Air Temperature Control

The intake air temperature control takes the form of a bi-metal flap valve mounted in the air stream to the air cleaner and senses the temperature of the incoming air. Cool air is admitted from the engine compartment, and hot air from a shroud around the exhaust manifold. At an air intake of 18°C or less, only hot air is admitted to the air cleaner. At a temperature of 33°C, only cool air is admitted.

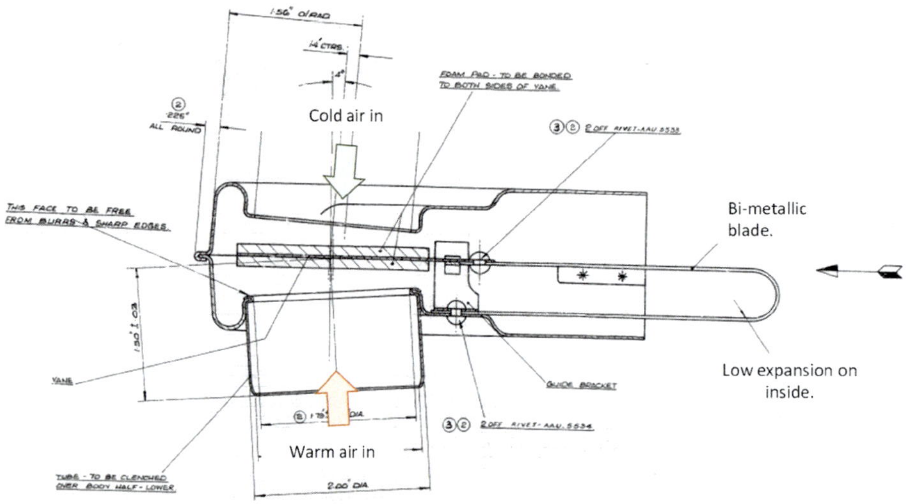

Fig. 3.1.3 *Intake air temperature control.*

3.1.4 Carburetter

The SU carburetter is similar to earlier type in its basic mode of operation but differs appreciably on account of emission control features.

Item	Feature	Purpose
Overrun limiting valve	Spring loaded plate in the throttle disc.	Under high manifold vacuum (overrun condition), valve opens allowing fuel air mixture through the throttle disc allowing normal combustion to occur.
Metering needle	Spring loaded needle	Light spring load against the jet so that the orifice is consistent.
Evaporative emission control	Float chamber vent	Allows evaporative emissions from float bowl to vent into the charcoal canister.
Crankcase emission control	Integral PCV	A constant depression is maintained in the crankcase by a take-off from the "constant depression" zone of the carburettor. The depression rises with engine load and speed. Crankcase gases are ducted through an oil separator and flame arrester on engine side cover. The crankcase depression is also used to purge the charcoal canister of adsorbed HC.

3.1.5 Air Injection

The AC Delco air pump is a vane type of operation driven by a belt from the crankshaft and draws air from the engine compartment, through a centrifugal filter, and then through a pressure limiting valve, diverter, and check valve to the air manifold.

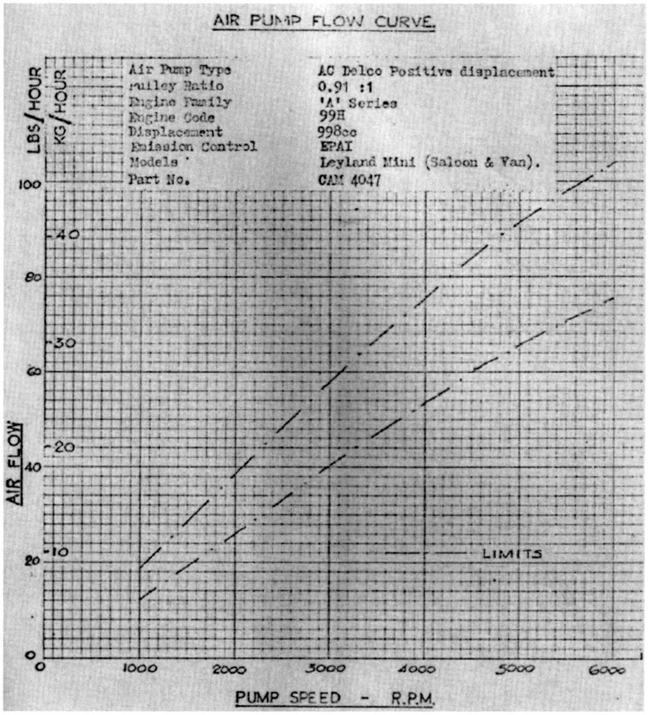

Fig. 3.1.4 *Air pump flow curve.*

Air injector pipes pass through the cylinder head and direct air to the back of the exhaust valve, where unburnt HC and CO can mix with oxygen and are then burnt.

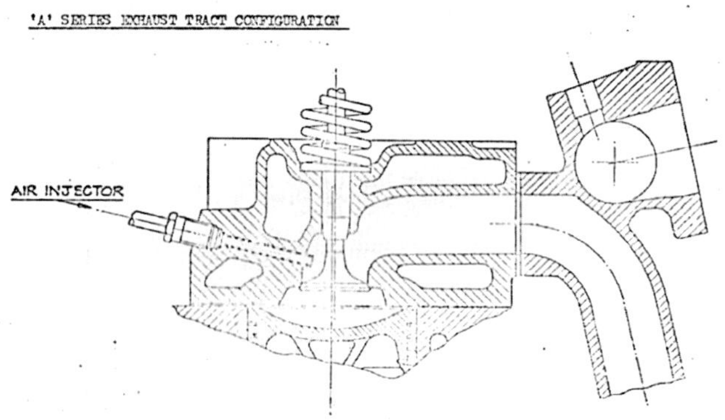

Fig. 3.1.5 *Exhaust port showing air injector.*

Fig. 3.1.4 shows the arrangement for the air injector, pointing slightly upstream to the exhaust gas flow through the exhaust valve.

Air pressure is dependent on engine speed and also exhaust back pressure. The pressure release valve prevents over-pressure, and a gulp valve reduces air to the injectors under rapid throttle closing to prevent back-firing. The gulp valve diverts air into the inlet manifold where it weakens the temporarily rich mixture thus allowing normal combustion. The check valve protects the air pump from reverse flow in case of pump failure (e.g. the drive belt fails). The diverter valve diverts air to atmosphere when the choke is in operation. The diverter valve is operated mechanically (via the choke cable) or by a vacuum switch which senses the main jet drop when the choke is pulled out.

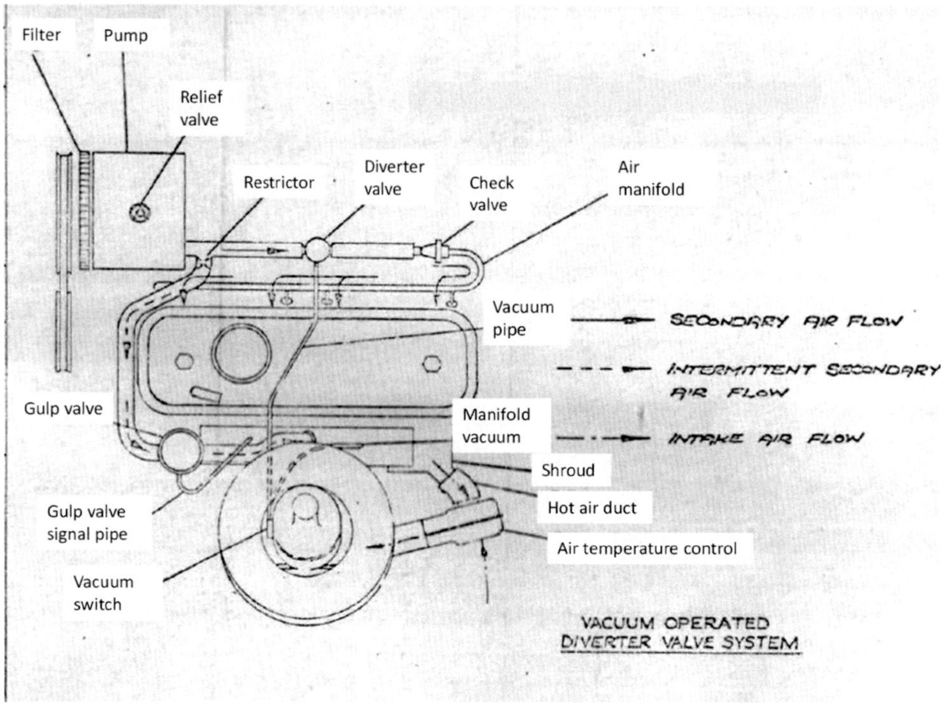

Fig. 3.1.6 *Exhaust port air injection general layout.*

3.1.6 Evaporative Loss Control (ELC)

The carburetter float bowl and fuel tank are vented to atmosphere through a canister filled with activated charcoal. The canister is fitted in the engine compartment. Vent and purge pipes are connected to the space above the charcoal material, and the lower space is exposed to atmosphere. When the engine is not running, hydrocarbon vapour is adsorbed on to the surface of the charcoal. When the engine is running, crankcase depression draws in air from the atmosphere in the reverse direction and purges the charcoal of adsorbed vapours. The vapour is drawn directly into the rocker cover area and then burnt via the PCV system.

The fuel tank for the van is fitted with a limited fill chamber so that thermal expansion of liquid fuel does not result in liquid reaching the charcoal canister. The position of the filler neck for the saloon model is such that it is not possible to over fill the tank and no modifications are required, except the use of a sealed fuel tank cap.

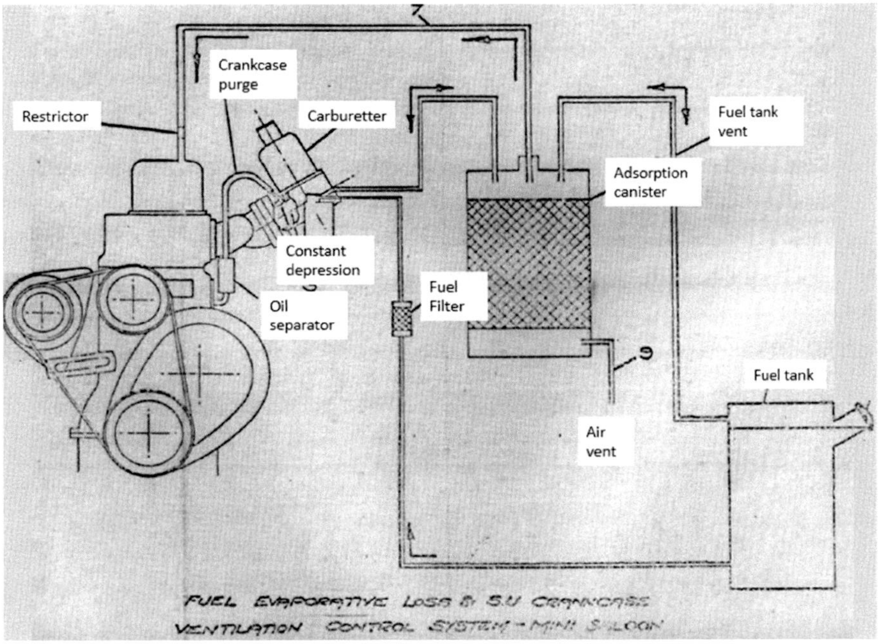

Fig. 3.1.7 *Evaporative emissions charcoal canister.*

The fuel cap is fitted with a vacuum relief and pressure relief valves.

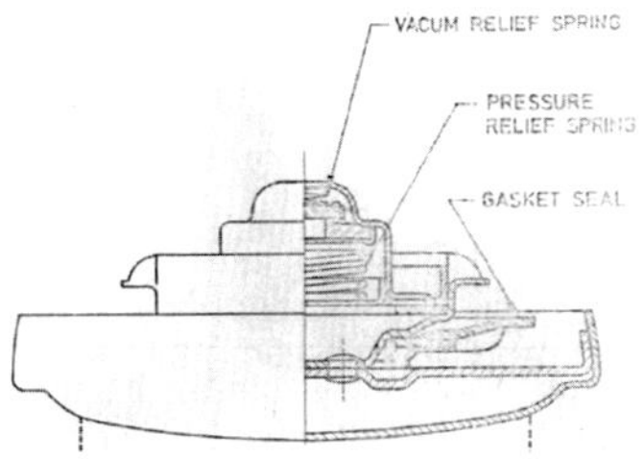

Fig. 3.1.8 *Fuel tank for evaporative emission control.*

3.2 Emission Control Testing

ADR27A testing was performed at two sites at Leyland Australia. An emission control testing laboratory was established near the Product Engineering building at Zetland in 1972. This facility kept operating until the factory closed in 1974 but was retained and operated by the Commonwealth Department of Transport until 1985 for research into the introduction of unleaded fuel in Australia.

A second laboratory was built at Moorebank, next door to the Parts and Accessories division. This was a large purpose built Environmental Centre, but with the wind-down of operations after 1974, it became a reduced facility for emissions testing and was used for product engineering development and then later, manufacture, of the Army Land Rover models.

An emissions test is performed on a vehicle which has been parked beforehand for a prescribed period (usually overnight) at a constant temperature. The vehicle is then positioned on a chassis dynamometer, but one that is fitted with inertia flywheels whose moment of inertia can be changed to suit the mass of the vehicle being tested. The dynamometer brake applies calibrated wind and rolling resistance, while the inertia wheels allow the engine to apply tractive effort necessary for acceleration.

The driver operates the vehicle in the normal manner. Vehicle speed is measured electronically, and a visual display plotted against time is shown. The driver has to operate the vehicle so that the speed matches the prescribed periods of idle, acceleration, constant velocity, and deceleration. The driving schedule is designed to simulate a "city cycle" and a separate higher speed schedule, a "highway cycle. A city and highway cycle performed back-to-back test would take about 40 minutes.

Fig. 3.2.1 *Chassis dynamometer with flywheels, water brake for rolling and air resistance. Driver operates the vehicle according to a strip chart of velocity vs time visible out the driver's window. (Zetland laboratory)*

A proportion of the exhaust gases are continually drawn off and then analysed for the concentrations of CO, HC, CO_2 and NO_x. The gas analysis subtracts the measured amounts for the concentration of these gases in the background air.

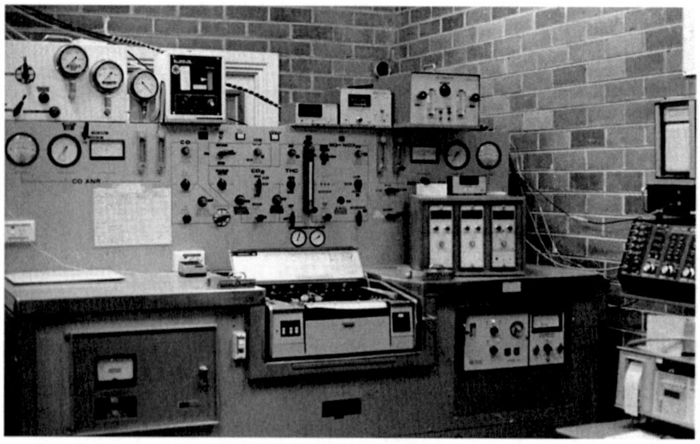

Fig. 3.2.2 *Gas analysis console. Exhaust gas concentrations for CO, CO2, HC and NOx are measured and expressed in grams/kilometre for comparison with legislated limits. (Zetland laboratory)*

Fig. 3.2.3 shows the plumbing and layout diagram for the exhaust gas analysis console. Calibration gases, background air, and the diluted vehicle exhaust can be fed through each analyser in the proper sequence. At the time, the gas concentrations were calculated using a time-sharing computer connected via a telephone.

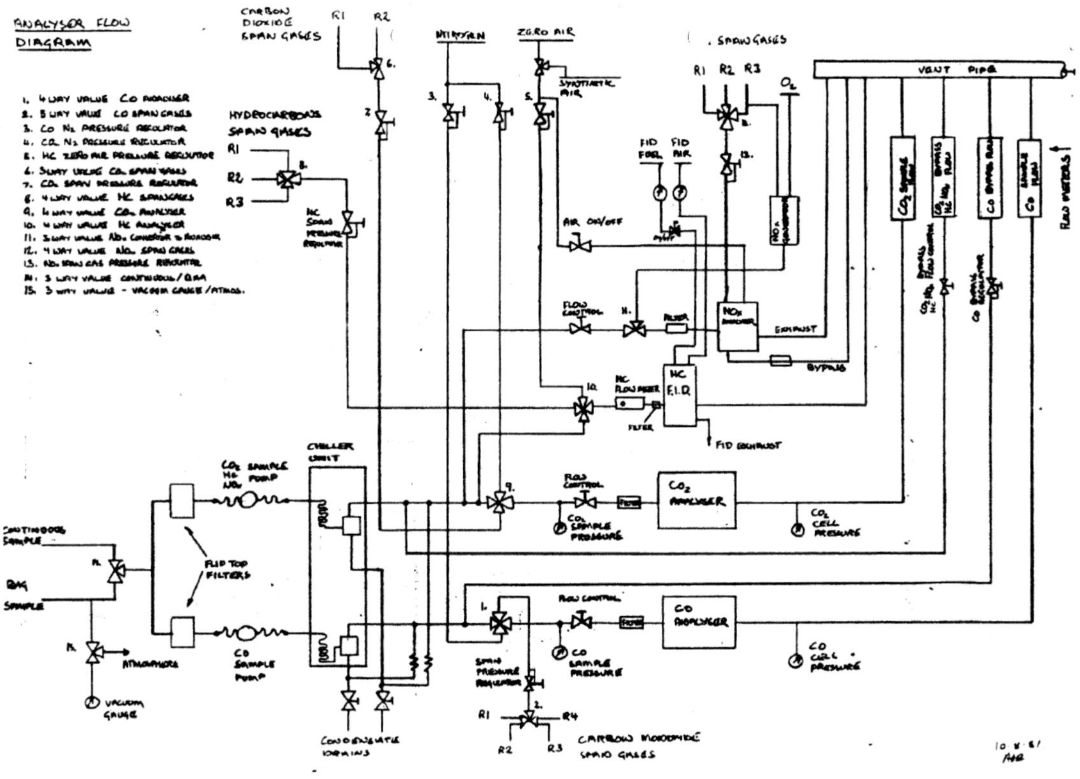

Fig. 3.2.3 *Schematic of exhaust analysis system. (Hand drawn by the present author in 1981).*

A by-product of the exhaust emissions analysis is a computation of fuel consumption since, if the chemical composition of the fuel is known, then the concentration of carbon in the CO, CO_2 and HC in the exhaust reflects the fuel consumed.

CHAPTER 4. IGNITION SYSTEM

4.1 Introduction

The process of igniting the fuel air mixture in the combustion chamber at the right time, and in the right cylinder, is the province of the ignition system. The ignition system must first generate a high enough voltage to cause a spark at the spark plug, and the distributor must distribute this high voltage so that the correct spark plug fires.

The voltage at which a spark appears at the spark plug depends upon several factors. The most important is the shape of the electrodes. The sharper the edges of the electrodes, the lower the sparking voltage needs to be. This is because an electric field is more dense around the space at a sharp corner compared to a rounded corner, and it is the strength of the electric field that causes ionization of the gases in this vicinity, and hence the establishment of a spark. The ideal shape of the spark plug electrode would be a sharp pin, but such a geometry would be subject to erosion by heat.

The larger the gap, the higher the voltage required to initiate a spark. The presence of fuel in the mixture slightly lowers the sparking voltage. The higher the pressure, the higher the sparking voltage. The lower the temperature, the lower the sparking voltage. The rate of application of the high voltage also affects the sparking voltage, a higher rate resulting in the sparking voltage increasing.

The spark itself is thought to consist of two components. The first is called the capacitive component and coincides with the opening of the points in the distributor. This is a bright, short flash at the spark plug. This is immediately followed by what is called the inductive component which has the appearance of a short series of fainter sparks over a relatively longer period. It is the capacitive component of the spark that initiates ignition of the mixture. It is the energy of the capacitive component, which depends greatly on the rate of application of voltage, that is the most important factor in producing ignition. What happens thereafter at the spark plug is of little consequence.

4.2 The Ignition Coil

The standard ignition system used on nearly all BMC/Leyland cars consists of points, condenser, coil, distributor, and spark plugs. This highly evolved system was preceded by trembler coil and hot wire methods. As is now well known, it is the switching action of the mechanical points that induces a high voltage in the ignition coil which is sufficient to create a spark at the spark plug inside the cylinder. The distributor itself distributes this high voltage to the correct spark plug at the correct time in the compression stroke.

However, there are some details about this process that are worth knowing in case of problems and modifications.

4.2.1 Faraday's Law

Fig. 4.2.1 shows the general arrangement. The operation depends upon Faraday's law of induction. This law states that the induced voltage in a coil depends upon the rate of change of current through the coil and the inductance of the coil. When the points are closed, current flows from the battery into the primary side of the ignition coil. The rate of change of current during this process depends upon the resistance of the coil, the inductance of the coil, and the voltage applied to the coil.

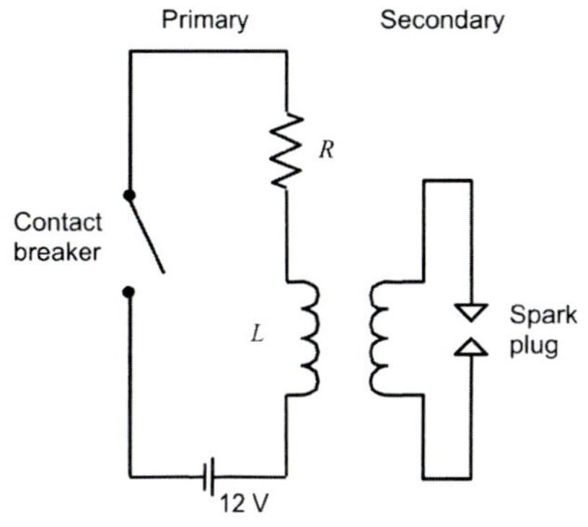

Fig. 4.2.1 *Schematic of ignition system*

There is an equation that describes how the current changes in the coil when the points close:

$$i = \frac{V}{R}\left(1 - e^{\frac{-Rt}{L}}\right)$$

In this equation, i is the primary current at some time t, V is the 12 V applied to the coil, R is the resistance of the primary, L is the inductance of the primary side of the coil.

What this equation tells us is that the current increases over time as shown in Fig. 4.2.2. I is the final current reached after a relatively long time.

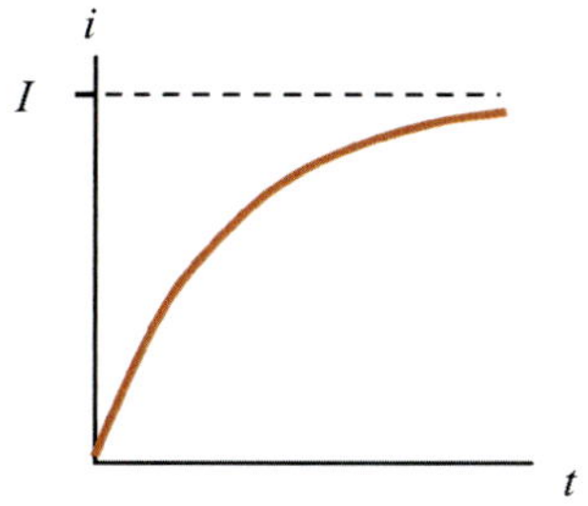

Fig. 4.2.2 *Current vs time in the primary side of the ignition coil.*

The time for the initial rate of increase in current is called the "time constant". It is desirable that the time constant be fairly short because the points are only closed for a brief time, and if the rate of current increase is too slow, then there will be insufficient build-up of current flowing in the primary during the time available that the points are closed. As the rpm of our engine increases, the rate of current rise in the primary has to be even shorter to take into account the even less time that the points remain closed.

Consider a 4 cylinder engine which has a specified dwell angle of say 60°. That is, this is the angle of rotation of the distributor shaft over which the points are closed. At a speed of 850 rpm, the engine requires about 28 sparks per second to operate. Calculations show that the points are closed for 25 msec for each spark. Using the equation above, this corresponds to a current through the points of 8A, or, in this case, about the DC value. This is called the "break current". At 4200 rpm, the spark rate is 140 per second, the points are closed for 5 msec, and the "break current" is 5.6A. In practice, a minimum of about 2 msec dwell time is required, corresponding to an engine speed of 8500 rpm and a break current of 3.6A.

These calculations assume that the heel of the contact breaker remains in contact with the distributor cam. It is desirable therefore that the cam has a somewhat gentle slope, but if this is done, then there is an undesirable variation in ignition timing with only a small change in points gap which occurs naturally as a result of wear during service.

A typical Lucas coil has a DC resistance of about 2 – 3Ω and an inductance of about 5mH.

Interestingly, the time constant τ for an *RL* circuit (the time at the initial rate of increase in primary current) is given by:

$$\tau = \frac{L}{R}$$

This gives a measure of the response time of an RL circuit. Here, *R* is the resistance of the primary circuit (including that of the coil winding itself) and *L* is the inductance of the primary coil.

When the points close, as would be expected, the greater the inductance, the longer the response time since the back emf is higher, and so requires a longer time to dissipate. What might not seem so obvious is the effect of the resistance.

If the resistance is larger, then the time constant is reduced. What this means is that (if we imagine the resistive part of the circuit separately to the inductive part) the voltage across the resistance rises more quickly for a lower resistance, the back emf across the inductor falls more quickly as a result, but the current tops out at a lower value of *I*. If, in a particular vehicle, the primary resistance is thus too high (such as might occur at the points), then there might not be sufficient energy stored in the coil to induce a spark at the spark plug and a misfire may result.

4.2.2 Generating a High Voltage

When the ignition points close, current flows into the primary side of the coil. The primary, being an inductor, opposes this current via the formation of a "back emf" according to Lenz's law. The magnitude of the back emf depends upon the inductance and the dc resistance of the coil, both of which determine the rate of change of current. The result is that the current flow through the primary starts off at a low value (since the rate of change of current is a maximum as the beginning of the process) and rises to a steady state value which ultimately depends upon the dc resistance of the primary coil. The energy of the primary current becomes stored in the magnetic field of the primary within which the secondary coil is also located as it is wrapped around the primary. A high voltage is generated in the secondary coil at this point, but not sufficient to cross the spark plug gap. The ratio of turns of primary to secondary is of the order of 100, and so for a 12 volt system, the induced secondary voltage at the time the points close is only about 100V.

When the coil is fully "charged", the rate of change of current is zero and there is a dc current of several amps flowing through the primary. When the engine is running, this condition is momentary due to the dwell angle of the points but may be continuous if the ignition is turned on, the points closed, and the engine stalled. At this point the current is flowing from the SW terminal into the coil and to the CB terminal to earth (assuming "conventional current flow" and a negative earth system).

When the points open, the current is abruptly interrupted. The magnetic field thus changes (collapses), and since a changing magnetic field induces a current in the coil the induced emf tends to want to keep the primary current flowing (Lenz's law again). The rate of change of current is very high (because the points open suddenly, and there is no "resistor" in the circuit as there was when the points closed). This rate of change of current is opposite in sign compared to the rate of change of current when the coil was being charged. That is to say, a falling current rather than a rising current.

When the points close, the back emf opposes the 12 V at the SW terminal in a gradually decreasing manner. When the points open, the rate of change of current in the primary coil is very high. The "forward" emf is of the order of 200 to 300V and appears at the CB terminal of the coil – which is now no longer connected to earth. The forward emf, sometimes called the "kick" voltage, or the "primary kick", is responsible for a far higher voltage generated in the secondary compared to the case when the points were closing.

We would expect such a large kick voltage to result in arcing of the points. This arcing would act as a kind of resistor (current flows through the arc) and would thus decrease the rate of change of current and limit the magnitude of the secondary voltage. The capacitor is charged by the kick voltage, and this takes time. Time enough for the points to separate so that arcing across the points is much reduced. Because of the capacitor stores the energy which would otherwise be wasted at an arc at the points, the primary current is more abruptly terminated resulting in a larger kick and higher secondary voltage.

Initially, the capacitor is charged with the kick voltage, but when the voltage at the CB terminal reduces (as the magnetic field in the primary becomes weaker as it discharged), the capacitor then discharges current back into the primary (reverse direction to that which occurs when the coil is being "charged" by the voltage at the SW terminal), and as a result, there is a back emf opposing this current and the voltage at the CB terminal rises. This back emf dissipates as the capacitor is discharged, and then the coil now with a new magnetic field (of reduced intensity) discharges again. The result is a series of oscillations at the CB terminal of a diminishing nature and the spark (now established) continues until the oscillations are reduced along the "spark line", or "burn line".

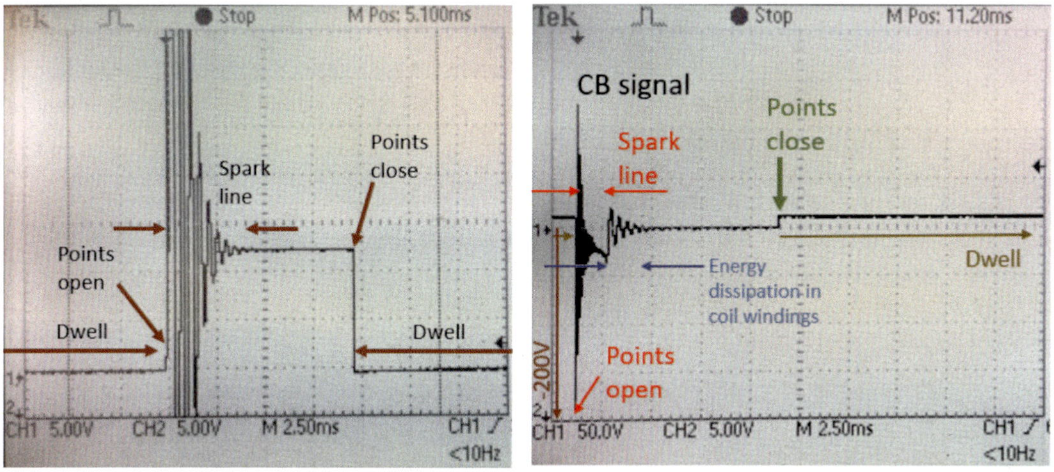

Fig. 4.2.3 *Voltage waveform at the CB terminal while engine is running. (a) Negative earth configuration (b) Positive earth.*

As far as igniting the fuel air mixture is concerned, it is the spark at the kick voltage that is the most important. The remaining spark during the spark line does not contribute significantly to combustion of the fuel.

It requires more voltage to initiate a spark at the spark plug than to maintain it since the gases in the spark plug gap have first to be ionised. Once ionised, the gases under dielectric breakdown and the voltage needed to maintain the spark is much reduced. The spark continues until there is no more ionised hydrocarbon gas present to sustain the spark current. The spark at the plug gap continues while the magnetic field diminishes and the oscillations due to the capacitor are reduced. The energy in the magnetic field appears as heat at the discharge at the spark plug gap. After the spark is extinguished, the remaining magnetic field dissipates and the potential at the CB terminal (and thus placed also across the capacitor) thus reduces. When the points finally close, a back emf is produced at the primary SW terminal opposing the battery voltage and the potential at the CB terminal goes negative before slowly rising to zero.

All this occurs over a small fraction of a second but can easily be seen on an oscilloscope. But a warning. An ordinary laboratory oscilloscope is usually not set up to handle hundreds of volts (the kick voltage) even though the coil might be powered by 12V, an automotive diagnostic oscilloscope which has the necessary capability to handle this input has to be used.

4.2.3 Ballast Resistor

BMC cars (in Australia) are usually not fitted with ballast resistors that are seen on some inferior makes, but it is worth knowing what they are for. The ballast resistor is placed in series with the primary side of the coil. Its resistance changes with temperature (doubling or tripling its resistance from cold to hot). The resistor is often bypassed by the cranking circuit. What this ballast resistor does is to make the process of ignition more reliable when cranking a cold engine. In this system, the ignition coil is designed to work on about 8 to 10V. When the starter motor is engaged, the battery voltage drops to about 10 V, and so this is applied directly to the coil. The coil is designed to have an acceptable rate of current rise when 10V is applied to its primary. It does this by having a lower primary resistance in its windings. When cranking stops and the engine is running, the primary current (now passing through the ballast resistor) causes the resistor to heat up and become more resistive, and so the 12 V gets lowered to about 10 V or so for the coil. Thus, the ignition coil operates at about 10 V all the time.

If a coil designed for a ballast resistor is used without the resistor, the 12 V battery voltage will be applied directly across the primary winding and it will become very hot, draw a lot of current, and burn out the points in short order.

4.2.4 Condenser

The condenser, or capacitor, absorbs the electric shock of the points opening and closing. Without the condenser, the points will burn out from arcing. If arcing occurs, the change in current in the primary will not be so rapid and the production of the required secondary voltage will be impaired – that is, the engine will misfire or just not run at all.

4.2.5 Coil

Experiments show that the most efficient spark occurs when the centre electrode is negative in polarity with respect to the engine block. The polarity of the spark depends upon the polarity of the voltage applied to the primary terminals. So, it is important (but not absolutely necessary) that the terminals marked "CB" and "SW" (for contact breaker and ignition switch) are correctly connected.

This applies even for a negative earth car. The centre electrode should be made negative with respect to the chassis. Since the secondary voltage (thousands of volts) is far greater than the vehicle's electrical system (12V), then having the centre electrode more negative than the chassis isn't a problem.

The coil is filled with oil. The process of induction and current flow in both primary and secondary is not 100% efficient. That is, considerable heat is generated in the coil when it is operating. The oil serves to insulate the windings, provide a method of cooling, and also, assists in contributing to the inductance of the device by virtue of its electrical properties (what is called the oil's permittivity).

The Lucas coil has a primary resistance of 2 – 3 Ω at 20°C.

4.2.6 Contact breaker

The action of the contact breaker is essential for good running. The contact breaker has to open and close at the right time, and remain closed for a long enough time for current to build up in the primary, open smartly so that a high voltage is induced in the secondary, and all at high speed. Mechanically, this depends on the profile of the cam lobes in the distributor. The shape of this cam is a compromise between good electrical performance and acceptable insensitivity to variations in contact gap caused by mechanical wear and user skill in setting the gap. In short, one must adhere to the manufacturer's dwell angle as closely as possible.

4.2.7 Electronic ignition

These days, electronic systems are the norm, and many BMC vehicle owners have fitted such systems to their cars. The simplest, and probably the easiest, is to replace the mechanical points and condenser with a solid state transistor or magnetic switch. This has the advantage of retaining the original exterior appearance of the distributor. An important point when fitting this system is to apply the white heat conduction compound (that should be supplied with the kit) to the base of the switch body so that heat generated inside the switch (by the large power transistor contained therein) will not overheat and fail.

Despite being more reliable, and able to be more accurately controlled by a computer, electronic systems still rely on Faraday to induce a spark at the spark plug and so the principles above are still relevant.

4.3 Spark Plug

4.3.1 Spark Plug Details

It is the opening of the points that interrupts the current in the ignition coil, and the resulting collapse of the magnetic field in the primary winding of the coil induces a high voltage in the secondary winding of the coil. This high voltage is applied to the electrodes of the spark plug whereupon the mixture in the cylinder is ignited. The high voltage is *distributed* to the correct plug via the rotation of the rotor button inside the distributor cap.

Now, firstly it is important to know that the spark at the plug is more efficient if the centre electrode is made negative with respect to the ground or cylinder head casting. This causes the spark to jump from the centre electrode to the outer edge of the plug. Even on a negative earth car, the polarity of the spark is more negative (by hundreds of volts) than the -12V at the outer ground terminal. This is why the terminals of the coil are marked + and -. Sometimes, the points side of the coil is marked CB for "contact breaker" and the other side SW for ignition switch. However, what is not generally appreciated is that there is a difference in internal connection for coils made for a positive earth car and a negative earth car. It has to do with the connection of the secondary winding inside the coil. When converting from positive earth to negative earth, one should not just swap the wires on the coil around, but one should obtain a coil wired for negative earth. The part numbers for coils for the two types of earthing systems are different.

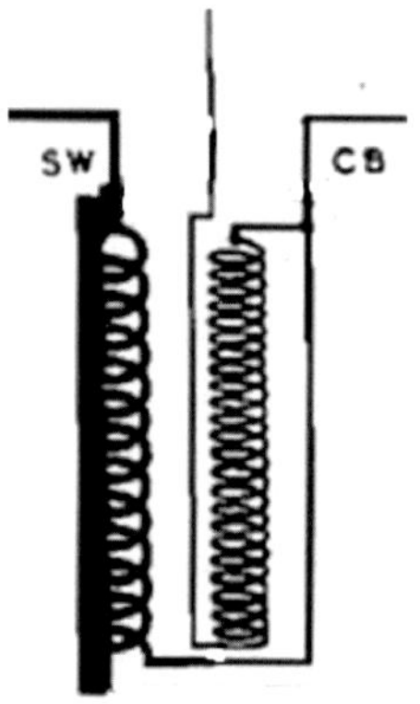

Fig. 4.3.1 Coil windings for negative earth car with SW being -VE and CB being +VE.

A further interesting point is that when the high voltage is distributed inside the cap, there is of course no direct connection between the rotor button and the cap terminals, so the high voltage has to arc across a small gap. Experiments show that this arcing causes metal to vapourise from the side from which the spark originates – that is, from the cap terminal (which is positive with respect to the rotor button) *to* the rotor button. Thus, when the coil polarity is correct, "wear" of the cap terminals will occur rather than wear of the rotor button – thus more evenly spreading out the wear of metal material rather than concentrating it at the rotor button.

A notable feature of the rotor button is the extension to the terminal in the direction of rotation. The width of the terminal has to be made long enough to allow the spark to proceed across the gap between it and the cap terminal while the rotor button is turning. If the rotor button terminal is too short, then it may pass by the cap terminal before all the energy is transferred – especially at high engine revs. Further, the extension in length of the rotor button is made in the direction of rotation. Lucas claim that this tends to minimise the chance of back-running of the engine (the spark is abruptly cut off) should the engine be turning in the opposite direction (which can sometimes happen during "running on" after the ignition is turned off.

Several types of spark plug are specified for the BMC range. The spark plug runs at about 2000°C at one end and at near ambient temperature on the other, over a length of a few centimetres. The transfer of heat from the plug tip to the cylinder head is important so that the tip does not melt, or become red-hot. If the plug tip becomes too hot, it may ignite the charge before the spark appears and detonation will result. If the plug tip is too cold, then the products of combustion may not burn off the tip and the plug will become fouled. The correct heat range of plug is therefore essential to good performance. The main difference between plugs of different heat ranges being the surface area of ceramic insulator material that joins the inner electrode to the outer metal thread – the more surface area the colder the plug.

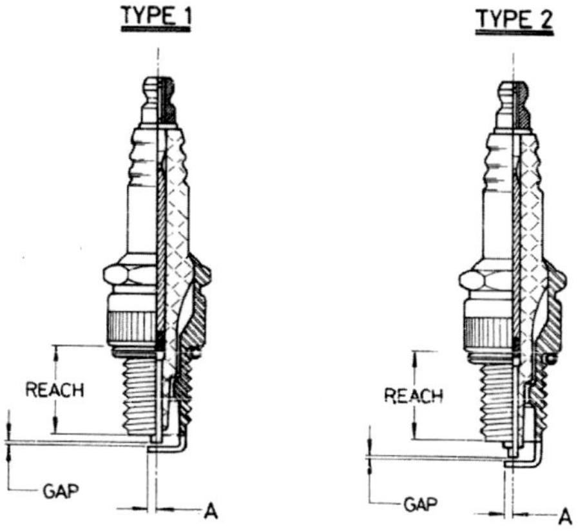

Fig. 4.3.2 *Spark plug types.*

Apart from the heat range, there are two general types which differ according to the protrusion of the centre electrode into the combustion chamber as shown below in Fig. 4.3.3.

PART NUMBER	TYPE	MAKERS INDENT	GAP (THOUS)	THREAD	REACH	MAKERS GASKET	A
AYG 2221	1	K.L.G. FE 75	24 to 26	14 M.M.	·75	690-06-07 690-05-02	·048±·005
AYG 0281	2	CHAMPION N6-Y	23 to 28	14 M.M.	·75	41804-8	·015±·015
AYG 0282	2	CHAMPION N9-Y	23 to 28	14 M.M.	·75	41804-8	·050±·015
AYG 0283	1	CHAMPION N5	23 to 28	14 M.M.	·75	41804-8	·041±·015
AYK 0098	1	CHAMPION N8	24 to 26	14 M.M.	·75	41804-8	·041±·015
AYD 0206	1	CHAMPION L-10	23 to 28	14 M.M.	·50	FGN-8A	·041±·015
AYD 0362	2	CHAMPION L-87Y	23 to 28	14 M.M.	·472	FGN-8A	·050±·015

Fig. 4.3.3 *Spark plug data.*

On very early BMC engines, the spark plug leads are often found to be terminated with uninsulated metal elbows that attach to the top of the spark plugs. There are, however, several types of high tension leads and some are shown in Fig. 4.3.5. For those with an insulating terminal sleeve, what may not often be appreciated is that the sleeve is not supposed to touch the spark plug insulator at all. This gives the longest possible insulated path along a surface from the top terminal to the vehicle earth (which is why the ceramic part is ribbed) and if the sleeve were to be a tight fit, it would then become a much shorter path. The rubber sleeve does indeed protect the terminal (if it is pushed on correctly) but doesn't "short out" the ribbed insulator.

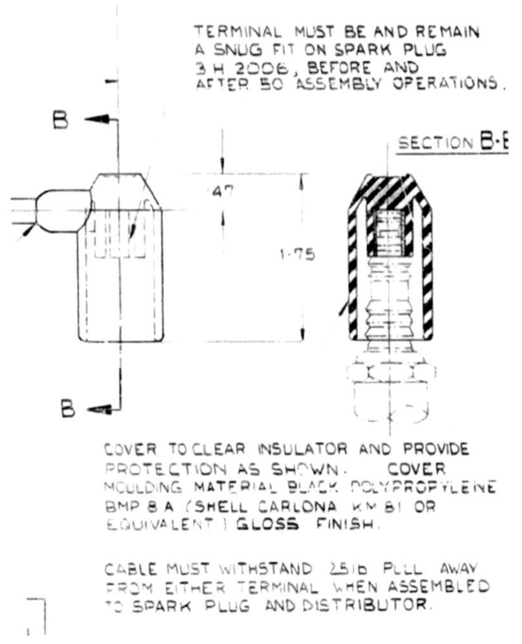

Fig. 4.3.4 *Spark plug cap.*

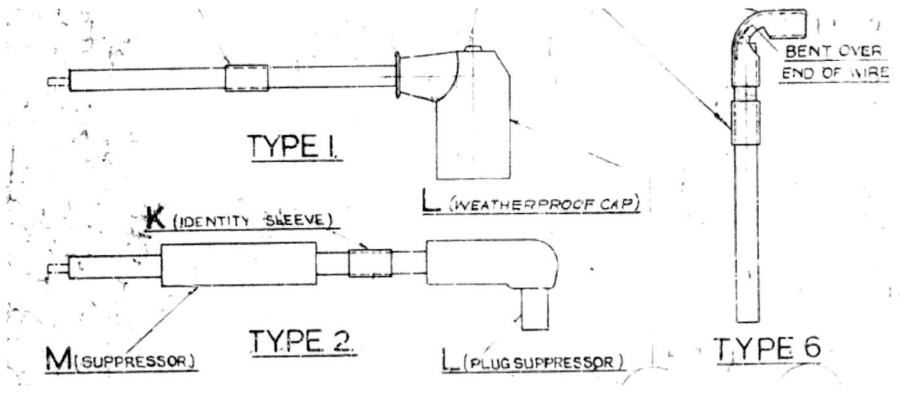

Fig. 4.3.5 *HV lead types and caps.*

4.3.2 Spark Plug Testing

Testing for the presence of a spark is one of the most common diagnostic procedures undertaken. One can test this by removing an ignition lead from a plug and holding it near to the cylinder head while an accomplice turns over the engine. Such a procedure may indeed show the presence of a spark – but is unsafe. The chance of the operator receiving an electric shock – which can be fatal in some electronic ignition systems, or causing a fire is substantial.

The correct tool for this job is a spark indicator. This is a small vacuum tube. The rarefied air inside the tube glows when a high voltage is applied to it. The end of this device is held near a plug lead or spark plug, and a glow will appear in the tube inside the plastic housing if there is high voltage present. The stronger the glow, the higher the voltage present. So, by running this device along the length of the HT leads, one can locate breaks and leaks in the lead, plus also estimate a comparison of firing voltage for each plug.

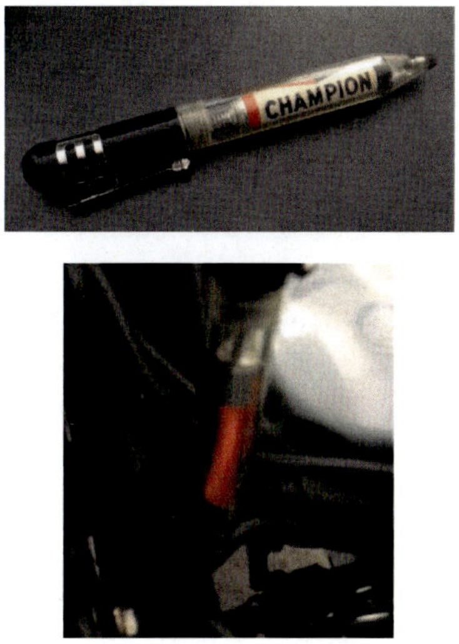

Fig. 4.3.5 *HV probe.*

A better measurement of firing voltage can be made with an oscilloscope – such as that found on trolley-mounted engine analysers. A portable version is available that is a practical alternative for a domestic situation.

Fig. 4.3.6 *Oscilloscope.*

With this device, an indication of firing voltage appears on a view screen in the end of the apparatus. It is powered from the vehicle battery, and a connection is made to the centre high voltage terminal of the cap – and another lead is clipped to the last plug lead in the firing order.

The resulting pattern shows an indication of firing voltage for each plug in firing order. A trace which falls above the red line indicates too high a firing voltage (gap too wide) while a very short pulse indicates too low a voltage (gap too narrow or some other short circuit).

4.4 Ignition Advance

The ignition system is responsible for igniting the fuel/air mixture in the combustion chamber. It may seem counter-intuitive that the ignition timing on the majority of cars is set so that the spark will occur *before* the piston reaches top dead centre (TDC). That is, the spark at the spark plug appears before the piston reaches the top of the compression stroke, and one might expect that the resulting explosion inside the combustion chamber would then serve to push the piston back down the way it came and either stop the crankshaft from turning, or forcing it to turn backwards.

Now, in order to explain why the spark is timed to occur before TDC, one must realise that the process of burning the fuel in the combustion chamber is not an instantaneous one. The flame front starts at the spark plug, and then over some microseconds, widens out and eventually reaches the furthest part of the combustion chamber. The ignition timing is set before TDC so that by the time the flame front has progressed through the combustion chamber, the maximum in the mean effective pressure (m.e.p.) occurs after TDC. The achievement of maximum m.e.p. after TDC also roughly coincides with the deliverance of maximum leverage to the crankshaft. The force from the pressure of combustion, multiplied by the length of the crank offset gives us torque to rotate the engine. When the connecting rod is at right angles to the crank, maximum leverage is attained.

The velocity of the flame front is assumed to be a constant (although it also varies with the amount of turbulence and mass, or charge, of air and fuel in the combustion chamber). When the engine speed increases, then the ignition timing must be advanced to compensate – because the maximum pressure still needs to occur at the same crankshaft angle after TDC when the leverage is optimum. So, the faster the engine speed, the more advanced the timing has to be so the flame front has enough time to create the maximum pressure when needed. It is the job of the mechanical advance to do this – increasing the ignition advance in proportion to engine speed. Note, the mechanical advance is in two stages – one of the weights comes into play at a certain rpm so that the advance curve is not linear throughout the rev range.

There is another advance mechanism often introduced and that is the vacuum advance. This advance mechanism provides additional advance, over and above that given by the mechanical advance, and is dependent on the load on the engine. The vacuum advance take-off port is not at inlet manifold vacuum, but usually from a place in the throttle body just above (on the atmospheric side) of the throttle plate. The maximum vacuum advance signal is obtained at part throttle at light load. Under these conditions, there is a small gap between the throttle plate and the carburetter body and the gas flow is high through this restriction, thus resulting in a depression in the vacuum advance line. Under heavy load, or acceleration, the vacuum advance signal drops off since the gas flow is not streamed past a small gap as is the case at light load.

At light throttle, with engine under light load, the charge (i.e. the mass of fuel and air) in the combustion chamber is much less than that at full throttle or under acceleration. Under these conditions, on the compression stroke, the compression pressure in the combustion chamber is less. The velocity of the flame front depends on pressure in the gas in the combustion chamber. At high compression pressures, the flame travels faster, while at low compression pressures, the flame travels slower. Therefore, at light throttle, the timing has to be advanced to account for the lower flame front velocity which occurs at lower compression pressures. At acceleration and heavy throttle, we have a greater charge entering the chamber, and lots of compression and so we do not need the extra advance – indeed, too much advance will cause many other problems.

It may be tempting to remove the vacuum advance since they are usually not seen on racing engines. For example, some Morris Cooper S distributors have no vacuum advance at all. Racing engines have no vacuum advance because for the most part, they are operated at wide open throttle and don't need to have any vacuum advance since it unlikely that they will be operated at light load conditions. It's best to take it off these engines so that the distributor advance curve has one less complication. For road use, it is better left on and working.[17]

[17] BMC Engineer Fulford wrote to UK to ask about the lack of advance on the Morris Cooper S engine and received a reply from J.R. Goffin of Morris Motors Engines Branch in Coventry whereupon Mr Goffin states "With regard to the fact that we do not fit a vacuum advance unit to the engine, this is because it is a sporting type of unit and it is not really necessary to fit one of these, in fact, if used for racing purposes this could become an embarrassment."

One might ask why there is not full vacuum advance at idle? The take-off port for the vacuum advance is usually on the atmospheric side (upper) of the throttle butterfly position. When motoring along at light throttle, the pressure at this port is very similar to manifold vacuum which is what we want in order to identify light load, part throttle conditions for maximum advance and full throttle low vacuum condition when advance is not required. At idle, when the port is closed off, the vacuum advance drops to zero. The reason for this is not so easily found – but is most likely done to produce a steady and smooth idle condition since at idle, the manifold vacuum is indeed very high, but it is also more likely to be pulsating. Better to have no vacuum advance than one which is being pulled on and off by a fluctuating vacuum signal. Also, the idle condition requires a significantly richer mixture (on old cars, about 4% carbon monoxide CO in the tailpipe emissions) compared to light running conditions (where the mixture is about 1 % CO) and this may also influence the choice of advance characteristics, especially for emission control vehicles.

Now, there is an interesting correlation between ignition timing and engine temperature. If the timing is too retarded, mean effective pressure is lost (because the piston is too far down on the power stroke before maximum m.e.p. is obtained) but a greater surface area of metal inside the cylinder is exposed to hot gases. So, more heat is transferred into the water jacket. As the throttle is opened more to compensate for loss of engine torque, thus increasing the amount of fuel burnt, more of the combustion gases are exposed to the cylinder walls, and so the engine runs hotter - sometimes to the point of overheating. Advancing the timing has the opposite effect, but then if there is too much advance, maximum m.e.p. occurs before that optimum leverage on the crankshaft and may also lead damaging detonation.

4.5 Distributor Testing

Broadly two types of distributors have been fitted to BMC engines of the 1960s. The Lucas 25D unit was fitted up to about September 1967 whereupon a change was made to the 29D locally made unit for the remainder. The 25D units usually have a side-entry cap while the 29D have a top-entry cap. There are many internal differences between the two.

The two most basic of distributor settings are of course the dwell angle (contact breaker gap) and the static ignition timing. The specified 60° dwell angle signifies that the distributor points are closed for 60° of distributor shaft rotation. This corresponds to a gap of about 0.015" at the points. This setting allows sufficient time for a magnetic field of sufficient strength to build in the ignition coil. When the points open, the magnetic field collapses, thus inducing a high voltage in the secondary winding of the coil and thus a spark at the spark plug. If the dwell angle is too low (gap too large), then the magnetic field in the coil will not have increased to the desired strength and a weak spark may eventuate. If the dwell angle is too long (gap too small) then the points may not separate cleanly and the resulting collapse of the magnetic field somewhat slowed down and again, a weak spark is the result.

The static ignition timing can be set either manually, with the engine at rest, by using an indicator, such as a test light, or an ohmmeter, to indicate points opening. The engine is turned by hand until the timing marks are aligned and then the distributor turned until the points are just opening. The rotation of the distributor is anti-clockwise when viewed from the top looking down into the insides. Thus, when adjusting the timing, it is best to approach the opening condition from closed to open, that is, turning the distributor body against the direction of rotation so as to take up any backlash rather than from open to closed. If there is not too much wear in the mechanical advance, then setting the timing this way can be done within one or two degrees compared to that using a timing light.

The use of a stroboscopic timing light is preferred over the static method since this is a dynamic setting that takes into account wear in the mechanism. It is important that the timing be adjusted with the engine rpm at a fairly low setting so that the mechanical advance is not activated, and with the vacuum advance unit disconnected.

Of course the ignition timing has to be adjusted depending on engine speed and load, and apart from making sure that the mechanical advance mechanism is not seized, and that the vacuum diaphragm is able to hold a vacuum without leaking, then there is little else that can be done by the average owner.

What would be most desirable is to measure the mechanical and vacuum advance throughout the rev range to make sure that the complete unit is working to the desired specifications. One of the most well-known instruments for this measurement is the Sun Distributor Tester.

In the 1950's to about the late 1980's Sun Electric Corporation made all manner of vehicle diagnostic equipment and is well known for its quality. The distributor tester dates from the 1940's and was still available new in the mid 1980's with various improvements along the way. One mounts the distributor in the clamps, engages the drive in the chuck, and connects the leads to the points and earth. Typically, there is a large degree wheel around the rim of the unit, and dwell, tacho and vacuum meters on the side. Some units also had a condenser test meter.

Fig. 4.5.1 *Sun distributor tester.*

When the unit is operating, a stroboscopic light indicates the points opening and this is shown on the degree wheel for all four firings (one rotation of the distributor shaft).

The rotation of the distributor can be adjusted using the hand wheel. The hand wheel essentially moves the position of a rubber wheel further up and down the surface of a cone drive underneath the drive plate. It is a very ingenious design since as the rubber wheel moves towards the larger diameter of the drive cone, it also moves into a smaller radius on the drive plate and so there are two "gear" mechanisms at play thus allowing the operator very precise control over the distributor rpm from zero up to about 3000 rpm.

A tachometer measures the rotational speed of the drive (which enables the unit to be used without a primary ignition circuit if desired), and a dwell meter is used to output of dwell, in degrees, and also doubles as a resistance measurement for the points. A separate pulse amplifier is used to test electronic distributors. There is a vacuum pump inside the unit which is connected to a gauge, and the level of vacuum can be adjusted with a control knob.

The tester allows a selection of clockwise or anticlockwise rotation. For Lucas distributors, anticlockwise is appropriate even though it might look that it is more natural for the cam to operate the points turning in the clockwise direction. Note, for safety, the distributor tester should be operated without the rotor button installed.

Any measurement needs to be made with reference to the factory recommended mechanical and vacuum advance curves. This information can be found in the workshop manual and hereupon there is some confusion. The information in the workshop manuals, the parts books and the service sheets is not consistent, probably due to the different advance curves specified for different countries to suit local fuel quality.

The distributor turns at half the crankshaft rpm. So, when the crankshaft makes one revolution, 360°, the distributor turns only 180°. This means that both the rpm and the advance, when we are talking about the distributor, are half of that of the crankshaft. Data in some manuals refer to *crankshaft* rpm and degrees. In the absence of any clear statement, it should be assumed that the specifications are *distributor* advance and rpm and not those of the crankshaft.

A graphical depiction of the advance curves is worth some scrutiny.

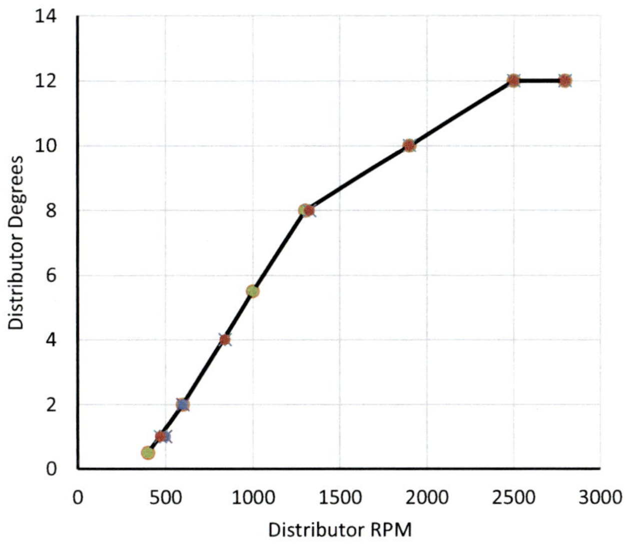

Fig. 4.5.2 *Centrifugal advance curves for various distributors for Austin 1800.*

Fig. 4.5.3 shows centrifugal advance specifications data for 25D and 29D distributors for the Austin 1800 ADO17 MKI and MKII range. The centrifugal advance is identical between these models. It is noted that the centrifugal advance curves are "bilinear". The first linear part (up to about 8 degrees) the centrifugal weights are controlled by the light spring. Beyond about 1300 rpm the heavy spring comes into play and the advance increases but at a slower rate until it tops out at about 12 degrees (distributor).

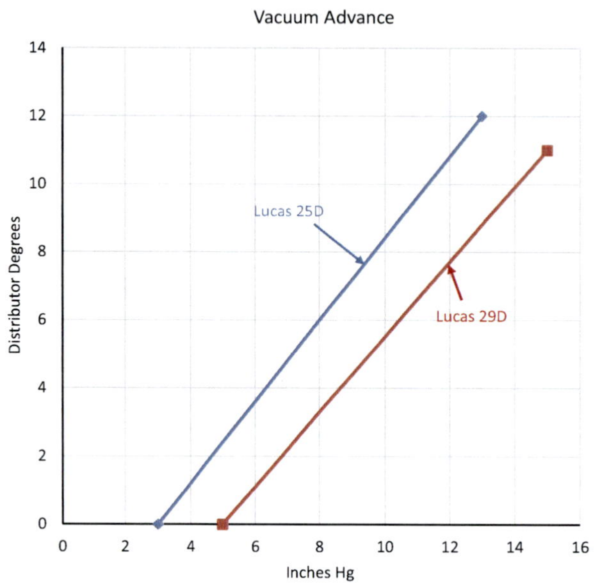

Fig. 4.5.3 *Vacuum advance, Austin 1800.*

For the vacuum advance, an Austin 1800 MK1 with Lucas 25D should begin advancing at 3" Hg reaching a maximum of 12 degrees (distributor) at 13" Hg while a MK2 with Lucas 29D begins at 5" Hg and tops out at 11 degrees at 15" Hg.

The above data are for the distributor. The advance actually experienced by the engine is the distributor data (converted to crankshaft rpm) added on to the static advance (i.e. 10° or 12°).

If the advance is measured using the distributor mounted in the engine and using the crankshaft pulley marks, the readings may be nothing like what is expected. For example, with the engine turning at 2800 rpm, this would be 1400 distributor rpm whereupon the advance (mechanical only) should be about 8° at the distributor. At the crankshaft, this would appear as 16° – which, when added to the initial 10° static advance, corresponds to a total of 26° at the crankshaft pulley. If the data in the workshop manual is interpreted as *crankshaft* readings, then at 2800 rpm one might expect to see 12+10=22° - and so it might be assumed something was wrong (4° retarded) when actually everything was correct.

One of the greatest advantages of the distributor tester is that the distributor is accessible. Apart from the advance curve measurement, certain preliminary tests are suggested by the Sun user manual.

Cam lobe accuracy can be measured by examining the variation in timing at each of the four strobe flashes on the dial. If there is a bent distributor shaft, for example, then the firing will be considerably uneven and the angular spacing of the indicator flashes will not be equal.

Breaker point alignment allows observation of the small spark at the contact points. This should be central to the points looking from above and from the side.

The dwell angle can be conveniently set with the distributor rotating and observing the dwell meter. Dwell variation with distributor speed can indicate worn shaft bushings.

The measurement of mechanical advance is easily done by varying the distributor rpm using the hand wheel and reading off the advance in degrees from one of the flashed timing marks. Measurements are made from the highest rpm down to the lowest as the worst case scenario should there be a sticking centrifugal weight.

In fig. 4.5.5, the data shows the readings before any adjustments or repairs were made, the factory specifications, and the readings are those after adjustment and repairs. As can be easily seen, the original readings were quite different from the factory specifications. This was tracked down to the fitment of an over-stiff primary spring in the centrifugal advance mechanism.

As well, there appeared to be an unacceptable amount of chatter from the mechanism below about 400 rpm (distributor). Close investigation revealed that there the primary spring was loose on the pivot in the relaxed position, causing the cam shaft to move a few degrees off the zero position without any resistance. The pressure of the points rubbing block was enough to cause the cam shaft to spring back and forwards at this free position causing the tang on the camshaft to knock against the centrifugal weight pivot post – thus causing a chattering sound, and as well, causing erratic timing at the same time. Since this movement took up some of the available rotational motion of the camshaft by the weights, the upper maximum centrifugal advance was not attainable.

There is little one can do to adjust springs on an individual basis, so adjustment of the centrifugal curve really comes down to a selection of different springs.

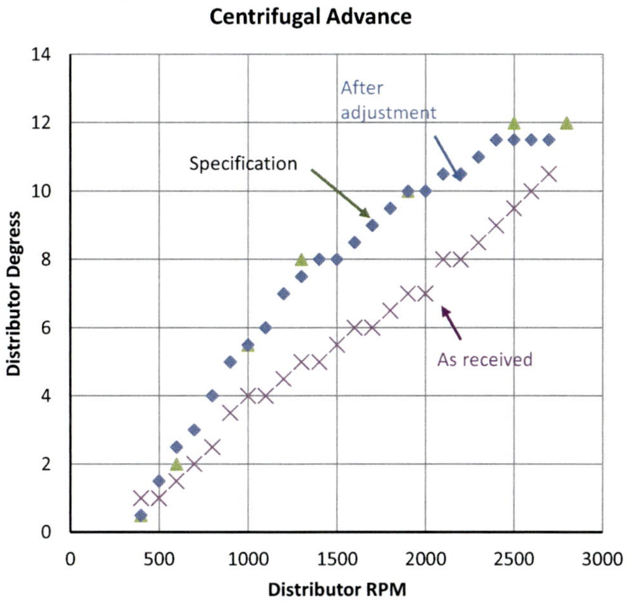

Fig. 4.5.4 *Comparison of specified, and actual centrifugal advance measurements in as-received condition and then after adjustment. Lucas 29D.*

CHAPTER 4. IGNITION SYSTEM 107

CHAPTER 5. COOLING SYSTEM

5.1 Introduction

The internal combustion engine, despite its great utility, has a thermal efficiency of only about 25% when one considers the energy available from combustion of the fuel and the mechanical work produced. This is far greater than previous attempts at mechanization and it can be appreciated arises from the high temperatures attained during combustion within the combustion chamber.

After combustion, the temperature in the cylinder is of the order of 1500°C which is sufficient to melt cast iron and aluminium. In practice, an engine may withstand such temperatures in the combustion chamber because they exist for a very short period, and heat is carried away by the cooling system.

As a further consideration, the viscosity of lubricating oil decreases with increasing temperature and so if the cylinder walls become too hot, the lubrication offered by the oil will diminish and friction thus increase, leading to seizure.

BMC factory documents describe the cooling system as pressurised thermos-syphon with pump and fan assistance. The thermo-syphon function relies upon a difference in pressure arising from the difference in density of the hot water in the water jackets and that of the cooler water in the radiator. The cooler water "sinks" and thus displaces hot water in the water jacket upwards into the top of the radiator, whereupon it is cooled and then recirculates.

In a pump circulation system, water enters the pump from the bottom radiator hose near the centre of the impeller and is then thrown outwards by centrifugal action imparted by rotating vanes. The cooled water, now under pressure, is directed into the water jacket of the cylinder block and then up to the cylinder head where it extracts heat and is then forced upwards under pump pressure through the thermostat into the top radiator hose.

A cooling fan is typically attached to the water pump, or electrically driven, to force air through the radiator.

5.2 Radiator

5.2.1 Radiator Function

The radiator serves to remove heat energy from the cooling water and so lower its temperature so that it may be recirculated for continuous cooling of the cylinder walls and cylinder head. The name is a misnomer in the sense that most of the heat transfer occurs via convection rather than radiation.

The radiator relies on there being a large surface area exposed to the water to be cooled, and a large surface area over which air flows. Heat is transferred to the air flowing through the radiator. Air may flow as a result of the forward motion of the vehicle, and also by the operation of the belt drive fan.

In a motor vehicle radiator, heat is transferred from the water by conduction to the tubes and fins of the radiator, and then by forced convection to the air which passes through the core.

Approximately 30% of the calorific value of the fuel is transferred to the cooling system. An additional 30% passes out through the exhaust gas, about 10% for mechanical inefficiencies such as friction and pumping losses, leaving about 20% for useful tractive effort.

The mechanical power output of a small four-cylinder engine is of the order of 40 kW, and so it can be expected that at full power, the cooling system will be required to dissipate heat at the rate of about 50kW. Some of this is done via radiation and convection direct from the engine assembly, but most of the heat has to be removed via the cooling system.

The temperature of the water entering the top of the radiator is typically about 85 °C.

The change in temperature of the water from the top to the bottom of the radiator can be computed by considering the loss of energy at a rate of say 50 kW via:

$$\dot{Q} = \dot{m}c(T_2 - T_1)$$

where $T_2 - T_1$ is the temperature difference between the top and bottom radiator tanks. The dot over the symbol Q indicates a heat flow rate, with units J/s or W. The parameter c is the specific heat of the water which can be taken to be 4.184 J/kg/K.

The radiator tubes are usually made of copper or brass which has a high heat conductivity. It is therefore the convective heat transfer from the fins to the air that is important.

The convective rate of heat flow is given by:

$$\dot{Q} = hA(T_2 - T_1)$$

h is the convective heat transfer coefficient, A is the surface area over which heat flow occurs, T_2 is the temperature of the water from which heat is to be extracted, and T_1 the temperature of the air flowing over the transfer surface.

It is immediately obvious from the equation above that the greater the surface area over which heat can be transferred increases the rate of heat flow. Also, the temperature different between the radiator tubes and the air passing through it also affects the rate of heat flow. Since the temperature of the water at the top of the radiator is higher than that at the bottom, then the rate of heat transfer is greater at the top of the radiator compared to the bottom. In an extreme case, using the example given previously, if the air passing through the radiator was at 65 °C, then there would be no heat transfer at the bottom of the radiator.

The convective heat transfer coefficient depends on the nature of the convecting fluid (in this case, air) and the nature of the heat exchanger geometry and construction. For air flowing

CHAPTER 5. COOLING SYSTEM 109

through a radiator at say 100 km/h, the convective heat transfer coefficient has a value of about 75 W/m^2/K. The lower the velocity of the air, the lower the convective heat transfer coefficient.

In the case of a motor vehicle radiator, the temperature of the water decreases from the top to the bottom of the radiator, and the temperature of the air increases from ambient as it passes over the fins to the exit. The effective temperature difference for the purposes of calculating convective heat flow in this situation is determined by what is called the log mean temperature difference and is given by:

$$\Delta T_m = \frac{(T_{Wout} - T_{Aout}) - (T_{Win} - T_{Ain})}{ln\big((T_{Wout} - T_{Aout})/(T_{Win} - T_{Ain})\big)}$$

In this equation, T_{win} is the top tank temperature of the water, T_{Wout} is the bottom tank temperature of the water, T_{Ain} is inlet air temperature, T_{Aout} is the outlet air temperature.

The difficulty with making quantitative analysis of an automotive cooling system is that there are a number of independent variables which have to be considered. The only fixed parameter is the top radiator tank temperature since this is set by the thermostat. The heat load to be dissipated depends on the throttle opening. An engine running at light throttle requires a lower heat transfer rate from the cooling system compared to the same engine running at full throttle. However, the water pump speed, and hence mass throughput of coolant through the system depends on engine rpm which may not necessarily be tied to throttle opening. The engine may operate at the same rpm with different throttle openings depending on road grade and other conditions. Further, the heat extracted through the radiator via convection depends on the temperature difference between the top and bottom tanks, the temperature difference between the radiator fins and the incoming air, and the velocity of the air passing through the fins which is affected by engine rpm via the fan and also road speed.

For motor vehicle cooling, the rate of heat flow out of the radiator has to be at least equal to the rate of heat energy uptake by the water as it passes through the engine. If the rate of heat flow is insufficient, such as when a clogged radiator is operated on a very hot day, overheating of the engine will result.

5.2.2 Radiators

5.2.2.1 Radiator Construction

As with most BMC front wheel drive vehicles, the radiator for the Morris Mini range is mounted sideways to the forward direction of the vehicle and so air entering the radiator is not fully front-on. There is a cooling fan fitted to the water pump hub and a close fitting shroud. The fan pushes air through the radiator core, unlike in most vehicles, where the fan draws air through the core. A tight fitting shroud can be used because the radiator is fixed to the engine, and not the body work, and so there is no real need for flexible couplings such as rubber radiator hoses, although of course these are fitted as a matter of course.

The cooling system has a capacity of about 3 litres. Since the density of water is about 1 kg/l, the mass of water in the system is about 3 kg.

In the Mini range, there are two grades of radiator. For the Morris 850, ADO15, the fins are spaced at 13 fins per inch, while larger capacities have 16 fins per inch. This increases the total surface area of the fins from about 1.6 m^2 to 1.9 m^2 within the same overall outer radiator dimensions.

At 16 horizontal fins per inch over a height of 11.44 ins (290 mm), each fin having a length of 10.5 ins (270 mm) and a width of 1.6 ins (40 mm), this corresponds to a total fin area ≈ 4 m^2 allowing for both sides of each fin. Counting up the 62 vertical tubes, three rows each of 290 mm height and 12.7 mm wide, the surface area is 0.46 m^2. The total convective area is thus 4.4 m^2.

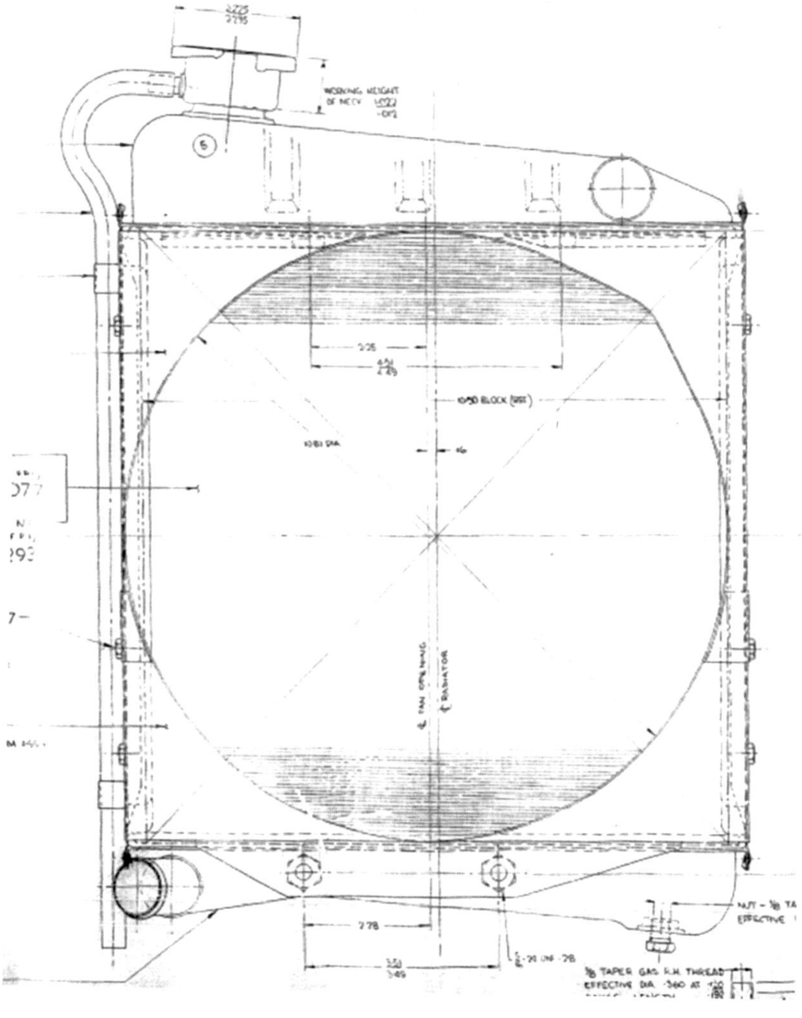

Fig. 5.2.1 *Radiator Morris Mini YDO5.*

5.2.2.2 Cooling System Performance

Despite the difficulties in analysis due to the many competing variables, it is possible to gain some feeling for the quantities involved based on some simple scenarios. In the examples shown here, the Morris 850, ADO15, engine and cooling system specifications are used to compared performance for three operating conditions: idle, light throttle, and full power.

The maximum available power from a typical BMC A series engine is approximately 30 kW at 5100 rpm. Since about 30% of the calorific value of the fuel may be dissipated by the cooling system, at an engine speed corresponding to maximum power, 5100 rpm, the cooling system may be called upon to dissipate about 40 kW. In this condition, the water pump is operating at maximum capacity due to the high engine speed, resulting in a mass flow rate of about 0.5 kg/s. The chances are that the vehicle is moving (if on level ground) at about 130 km/h leading to a fairly high heat transfer coefficient estimated at 75 $W/m^2/K$. It can be assumed that the upper radiator tank temperature is fixed at 85 °C by the operation of the thermostat.

Table 5.2.1 shows the estimated bottom tank temperature, and the required area *A* for forced convection based on the above, as well as estimates for air temperatures each side of the radiator. It is convenient to work in terms of convective area rather than estimating temperature using the measured area as an input due to the complications in extracting the temperatures from the log mean temperature difference.

At full throttle operation, the required area is about 16 m^2, but only 4 m^2 are available. This being the case, it is not advisable to operate the vehicle at full throttle for long periods unless the ambient temperature is much lower than the 25 °C used in these calculations.

Similar estimates can be made for light throttle and idle conditions, and the results are shown in Table 5.2.1.

Quantity	**Full throttle**	**Light Running**	**Idle**	**Unit**
Heat Capacity				
Mass flow rate	0.50	0.35	0.05	kg/sec
Specific heat	4.184	4.184	4.184	kJ/kg/K
T_2	85.00	85.00	85.00	°C top tank
Power	50.00	20.00	2.00	kW
T_1	61.10	71.34	75.44	°C bottom tank
ΔT_{water} top to bottom tanks	23.90	13.66	9.56	°C
Calculated Convection				
h	80	60	10	W/m2/K
T_{Win}	85	85	85	°C top tank
T_{Wout}	61.10	71.34	75.44	°C bottom tank
T_{Ain}	25	25	25	°C Air in
T_{Aout}	45	35	30	°C Air out
ΔT_{air}	33.37	47.19	52.38	°C LMTD
Convective Area	18.73	7.06	3.82	m^2

Table 5.2.1 *Calculations of cooling system performance Morris Mini.*

Considering the simplifying assumptions made, and also the effect of neglecting any heat flow by radiation and conduction, the results show that for this vehicle, the radiator is probably of sufficient capacity to provide safe operation under light throttle conditions and idle. At idle, the engine is simply turning against mechanical friction and the heat load used reflect this circumstance.

Experience shows that most problems arise when the engine is idling for long periods on account of the lack of air flow through the radiator. Adjusting the air flow by x2 in the above (by increasing *h*) decreases the require area for convection by a factor of 2.

These calculations service to show the effect of varying design parameters on cooling system performance. A manufacturer would test these assumptions by experiment and BMC's Experimental Department had facilities for these tests as shown in Fig. 5.2.2.

Fig. 5.2.2 *Cooling system test rig, BMC Experimental.*

5.3 Radiator Cap

5.3.1 Pressurized Cooling

For the most part, the cooling medium in BMC vehicles is water, and in some models, water and antifreeze. The boiling point of water depends upon atmospheric pressure. The change in boiling point as a function of pressure can be calculated using what is known as the Clausius-Clapeyron equation.

$$P \sim e^{-\frac{H}{RT}}$$

In this equation, H is the latent heat of vaporization expressed in units of J/mol, R is the universal gas constant 8.314 J/mol/K, P is the absolute pressure (the vapour pressure of the liquid), and T is the absolute temperature in degrees K.

To see how this equation applies to water in a vehicle's pressurized cooling system, we can plot the boiling point T as a function of pressure P using 100 °C at 101.3 kPa (standard atmospheric pressure) and T = 373K (100 °C) as a baseline. Taking the natural logarithm of both sides of the Clausius-Clapeyron equation and forming the ratio P_1/P_2, we obtain:

$$ln\left(\frac{P_1}{P_2}\right) = -\frac{H}{R}\left(\frac{1}{T_2} - \frac{1}{T_1}\right)$$

Where P_2 is the new vapour pressure, and T_2 is the new boiling point. Fig. 5.3.1 shows the variation in boiling point (T_2 in °C) as the pressure increases beyond that of standard atmospheric conditions of P_1 = 101.3 kPa (14.7 psi). Since cooling system pressures for vehicles are commonly expressed in psi, the data is shown here in terms of psi rather than metric units.

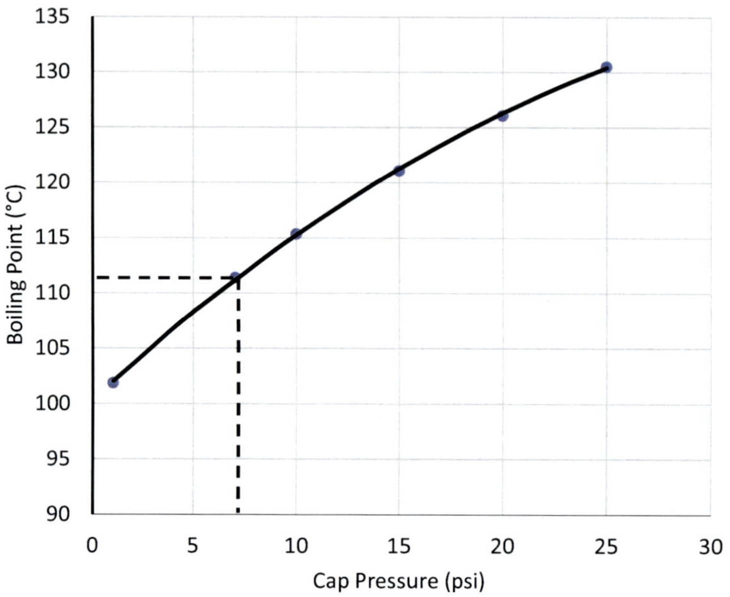

Fig. 5.3.1 *Boiling point vs Pressure in psi. At standard atmospheric pressure, the boiling point of water is 100°C. With a 7 psi radiator cap (dashed line), the boiling point is raised to about 112°C.*

The objective of the cooling system is to carry away waste heat generated by combustion of fuel. Relatively cool water from the radiator bottom hose flows through the block and the head and absorbs heat, and the heated water goes into the top of the radiator after which it makes its way down to the bottom then recirculates. Should the water get too hot, it may boil – which is particularly unfortunate, especially if the boiling is at some local hot spot. The transference of heat from metal to steam is significantly less efficient compared to metal to water and any boiling in the system is likely to result in overheated metal and undesirable distortion of the cylinder block or head. If the cooling system were not pressurized, boiling would occur at 100 °C which leaves little in reserve for operation on a hot day. Most engines are designed to operate with a pressurized cooling system with water at a temperature within the range 80 to 90°C.

5.3.2 Radiator Caps

Many radiator caps are common across in the BMC range of front wheel drive cars. For earlier models, radiators have a deep, or long neck. For later models with an expansion tank, the radiator cap with a short reach is required.

Table 5.3.1 shows some representative details.

Model	Cap	Pressure (psi)	Description
Mini			
Morris 850	AYA2059	7	Radiator (long neck)
Minor	AYA2059	7	Radiator (long neck)
Mini Deluxe	AYA2078	13	Radiator (long neck)
Mini 1100	AYA2078	13	Radiator (long neck)
Cooper	AYK2051	10	Radiator (long neck)
Cooper S	AYA2078	13	Radiator (long neck)
Morris 1100			
1100 and 1100S	AYB2037	13	Expansion Tank (Short neck)
1300	AYB2037	13	Expansion Tank (Short neck)
1500	AYB2037	13	Expansion Tank (Short neck)
Nomad	AYB2037	13	Expansion Tank (Short neck)
Austin 1800			
MKI	AYB2037	13	Expansion Tank (Short neck)
MKII Man	AYB2037	13	Expansion Tank (Short neck)
MKII Auto	AYH2486	14	Expansion Tank (Short neck)

Table 5.3.1 *Radiator cap/expansion cap details for selected models.*

The radiator cap is essentially two valves, a pressure relief valve and an often-forgotten vacuum relief valve. Pressure relief is in the form of compression of a large spring under the sealing disc. Vacuum relief is a lighter spring located on the other side of the pressure seal in the centre of the cap. Without vacuum relief, atmospheric pressure may case the thin tubes of the radiator core to collapse as the cooling system cools down to room temperature when the engine is turned off.

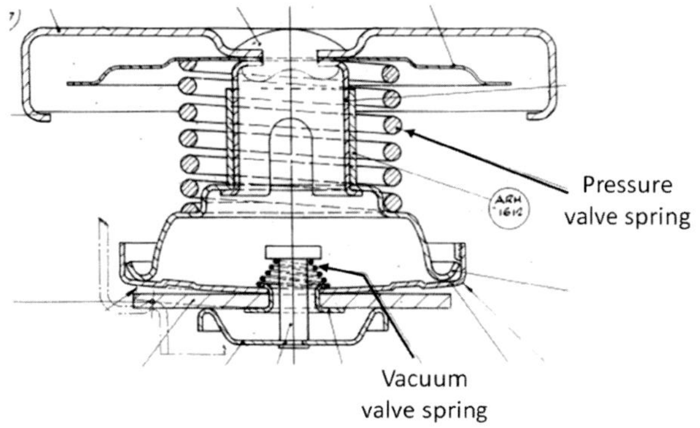

Fig. 5.3.2 *Radiator cap cross-section.*

5.4 Fan Belt

5.4.1 Belt Transmission

It is usual practice to find that the cooling fan for the radiator and the generator/alternator armatures are driven by pulleys via a belt from the crankshaft pulley. This simple, low cost, and relatively silent method of power transmission is worthy of detailed attention.

Fig. 5.4.1 shows a simplified arrangement for belt transmission. A flat belt is fitted over a flat pulley and the belt placed under tension. The application of a torque to the hub of the pulley via the belt relies on there being friction between the belt and the pulley. When there is no torque to be applied at the pulley hub, the tensions T_1 and T_2 are equal and there is no friction force. This is the static tension T_0.

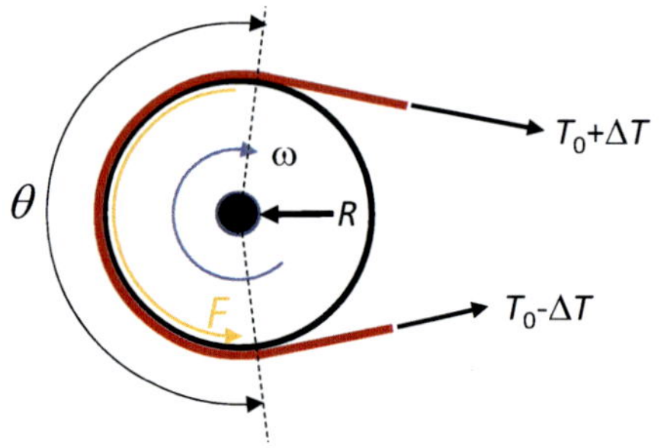

Fig. 5.4.1 *Flat belt over a pulley.*

When there is a mechanical torque to be to be applied to the hub, T_1 is increased in relation to T_2 by virtue of a distributed friction force between the belt and the pulley over the length of the contact. At the point of maximum adhesion (i.e. just before slip), the friction force at each point along the contacting surfaces is determined by the normal force at that point and the coefficient of friction.

Since in this condition, T_1 is greater than T_2, it can be appreciated at the tension in the belt must therefore change in some manner over the length of the contact between the belt and the pulley. As well, at the very beginning of the length of contact, the normal force is zero, and so is the friction force. At the middle of the length of contact, the normal force is a maximum, and thus so is the maximum friction force attainable before slipping. At the end of the length of contact, the normal force drops to zero and so does the friction force.

The relationship between the tension in each side of the belt, the coefficient of friction, and the angle over which the belt makes contact with the pulley is given by the Euler equation:

$$\frac{T_1}{T_2} = e^{\mu\theta}$$

In this equation, T_1 is the tension in the tight side of the belt, and T_2 is the tension in the loose side. μ is the coefficient of friction between the belt and the pulley, and θ (in radians) is the angle over which the belt makes contact with the pulley. This equation applies to the situation when the belt is just about to slip.

When the belt is a vee belt, the effective coefficient of friction is increased due to the wedging action of the belt in the groove and the equation becomes:

$$\frac{T_1}{T_2} = e^{\frac{\mu}{sin\beta}\theta}$$

Where β is the half angle at the bottom of the vee. This equation shows that the ratio of the tensions from the tight side to the loose side does not depend on the speed of rotation of the load, but only on the friction coefficient, the angle of the vee, and the angles over which contact is made. The overall tension may change depending on the load, but the ratio remains constant.

The power transmitted by the belt is the difference in the tensions multiplied by the linear velocity of the belt.

$$\begin{aligned} P &= (T_1 - T_2)v \\ &= T_1\left(1 - \frac{1}{e^{\frac{\mu}{\sin\beta}\theta}}\right)v \end{aligned}$$

The linear velocity *v* is related to the angular velocity ω and the radius *r* of the pulley by:

$$\begin{aligned} v &= r\omega \\ &= 2\pi r\frac{(rpm)}{3600} \end{aligned}$$

These equations show that the power transmission does depend on the tension in the belt (T_1) and the speed of rotation. T_2, at the moment of slipping, is in turn determined according to the Euler equation.

In most situations, the belt has an initial tension T_0. This being the case, when torque is to be transmitted, the tension in the tight side of the belt increases by an amount ΔT and that on the loose side by ΔT. Thus,

$$\begin{aligned} T_1 &= T_0 + \Delta T \\ T_2 &= T_0 - \Delta T \\ T_1 - T_0 &= T_0 - T_2 \end{aligned}$$

And

$$2T_0 = T_1 + T_2$$

An allowance can also be made for a reduction in tension arising from centrifugal forces resulting from the mass of the belt as it travels around the pulley.

5.4.2 Morris Mini Range

Figs. 5.4.1 and 5.4.2 show the general arrangement and dimensions, and Table 5.4.1 the design parameters for the Morris Mini range, YDO4, YDO5 and YDO6 as of 1971.

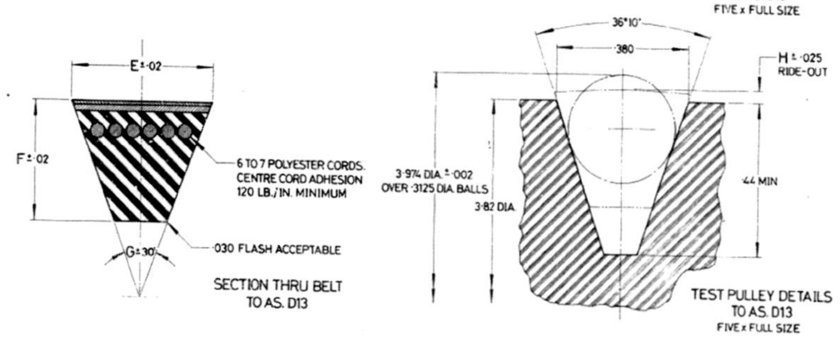

Fig. 5.4.2 *Fan belt details, Part No. AYG2271.*

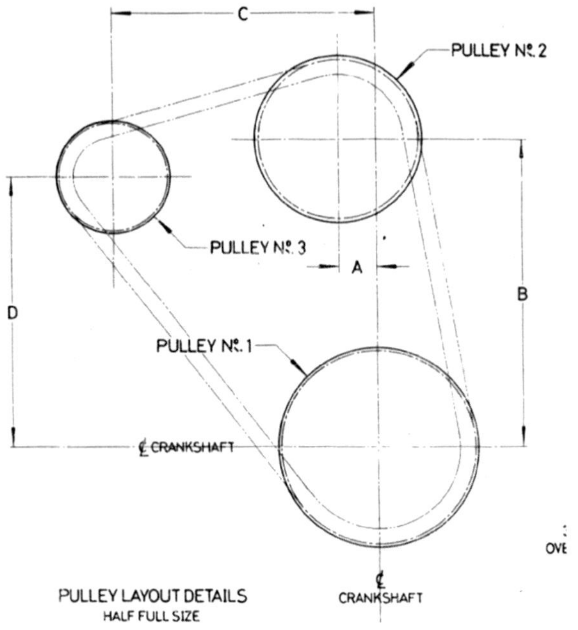

Fig. 5.4.3 *Fan belt arrangement for YDO4, YDO5 and YDO6.*

Fan Belt	Part No.		Pulley Details	C'shaft (No. 1)	Pump (No. 2)	Alternator (No. 3)	
Dimension	AYG2271	Unit	Effective Diameter	113.5	98.7	64.3	mm
A	21.3	mm	Contact Arc	152	95.5	112.5	deg
B	180.6	mm	Max RPM	6000	7265	10569	rpm
C	154.2	mm	Max Speed Change	54			revs/s²
D	158.8	mm	Max Power		2.42	1.61	kW
E	10.2	mm	Inertia		2.63E-03	3.22E-03	kg m²
F	8.6	mm					
G	40	deg					
H	1.0	mm					
L	260.4	mm					
Circumference	825.5	mm					
Tension	289.5	N					
Material	FBY-1A						

Table 5.4.1 *Fan belt and pulley details, YDO4, YDO5 and YDO6 as of 1971.*

While the dimensions and other specifications are more or less fixed by the factory specifications, the equations which describe belt friction allow quantitative comparison between different operating conditions. The only parameter that is under the control of the driver/operator is the belt tension.

Rearranging some of the previous equations show that the amount of power able to be transmitted depends linearly on the initial tension. For example, for YDO5 specifications above, at an initial tension of 290 N, and with a coefficient of friction of 0.3, and an angle of contact of 113.5°, the crankshaft pulley is capable of supplying about 15 kW before belt slippage, whereas if the initial tension were halved, the power available would also be halved. Given that the power to be transferred to the water pump and alternator is specified at 2.42 and 1.61 kW respectively, there appears to be plenty of crankshaft capacity in reserve to account for in-service wear and maladjustment.

BMC fan belts are designed to provide 25,000 miles service under normal running conditions with proper tension and alignment over a temperature range of -35°C to 71°C. Table 5.6.2 shows the design belt grades.

Grade	Description
FBY-1A	Premium quality for use with small pulley (under 2.8") diameters.
FBY-1B	Premium quality for use with pulleys 2.8" diameter and larger and for vehicles having power steering.
FBY-1C	Regular quality for use with pulleys 2.8" diameter and larger.

Table 5.4.2 *Grades of BMC V belts.*

5.5 Thermostat

5.5.1 Need for a Thermostat

It is well known that cars used for taxi work can accumulate high mileages without any engine work other than scheduled maintenance. The reason is that taxis and most commercial vehicles are in continuous use and thus avoid the deleterious conditions associated with a cold start of the engine.

Manufacturers set tolerances and clearances for pistons, rings and bearings, based on a certain normal operating temperature. This temperature is a compromise between being too hot (and therefore not enough cooling reserve from the radiator for hot days) and too cold (causing incomplete combustion and increased friction from the lubricating oil).

The most desirable scenario (apart from keeping engines heated electrically overnight) is to raise the temperature of the engine as quickly as possible and to keep it at its designed operating temperature. This is the function of the thermostat. By blocking or reducing the flow of cooling water during initial running, the temperature of the engine rises rapidly, and then, when the thermostat opens, it remains at the correct temperature. Excess fuel consumption from choke operation is reduced, and blow-by of harmful compounds to the crankcase is decreased, and what does blow by, is more readily boiled off in the sump and purged by the PCV system.

In the fully open condition, the size of the opening of the thermostat is taken into account by the manufacturer so that the correct flow rate of cooling water is achieved during normal operation.

5.5.2 Thermostat Design

Thermostats of older design are often unreliable. The mode of operation is based on the expansion of a bellows or concertina which is filled with acetone. Evaporation of the acetone (at about 60 to 80°C) expands the bellows and opens up the thermostat flap valve. When this type of thermostat fails, it remains open. Note also that this type of thermostat relies on a difference in pressure between the liquid inside the bellows and the water outside. This has some consequence for the use of a pressurised cooling system.

In later designs, thermostat operation is by way of expansion of a wax-filled chamber. The expansion of the wax forces a central pin upwards and opens the valve. If the wax fails to expand, then the thermostat remains closed. A jiggle pin allows air to pass through so that filling a cold system from empty can be done without bleeding the system from the top hose. Note that this type of thermostat is not pressure sensitive. The wax type thermostat is very reliable, and it is rare to encounter one that gives problems.

For the Mini range, early model Morris 850 vehicles were fitted with an 82°C Western Thomson thermostat of a bellows type 11G291 with an opening temperature of between 72°C and 76°C.

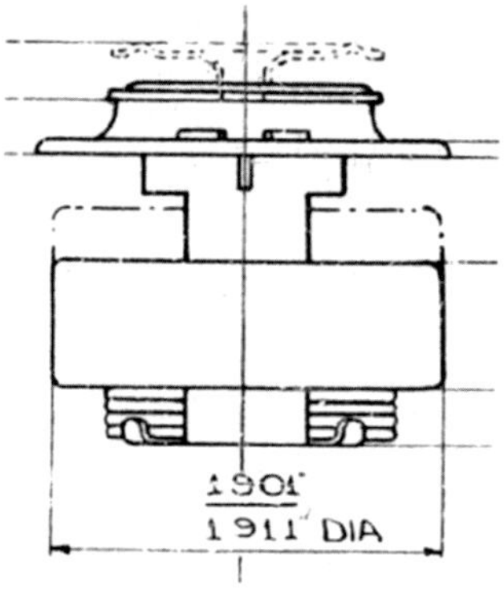

Fig. 5.5.1 *Bellows type thermostat.*

Later models are fitted with an 82°C wax type thermostat 13H956 introduced in 1964. A new thermostat, downward opening, was introduced in 1968.

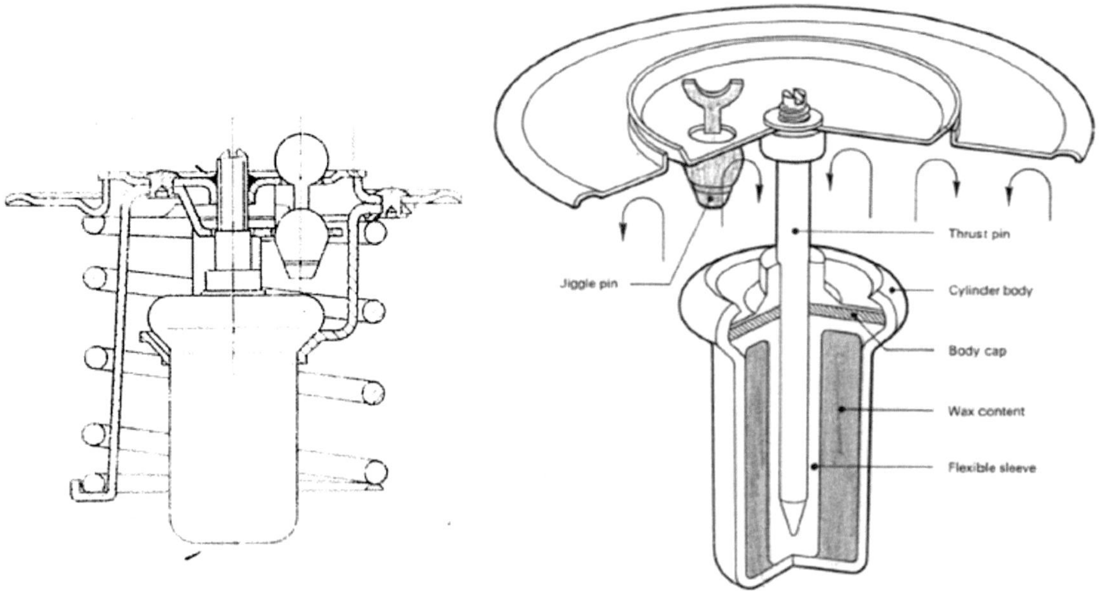

Fig. 5.5.2 *Wax type thermostat 13H956.*

CHAPTER 6. FUEL SYSTEM

6.1 Fluid Dynamics

Much of the operation of any carburetter is governed by the principles of fluid flow, both of gases and liquids.

When particles in a fluid follow smooth, well-defined paths, the flow is said to be laminar. Streamlines indicate the path of the particles. A vector drawn tangent to the streamline may be used to indicate velocity at any particular point.

At low particle velocities, in the absence of any disturbances, the flow of fluid in a pipe is usually laminar. However, as the velocity increases, irregularities in the walls of the pipe cause vortices to form and at higher velocities, the flow becomes turbulent.

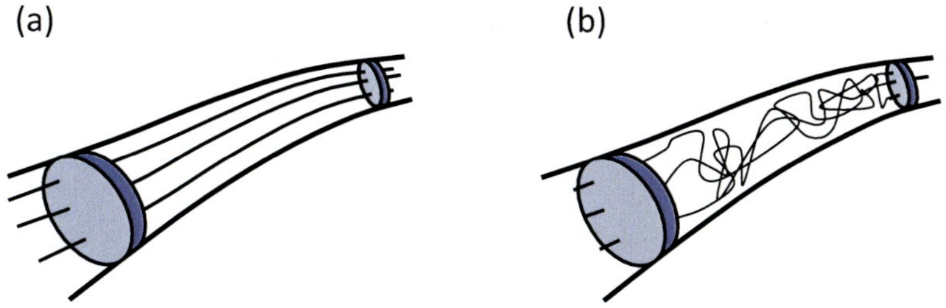

Fig. 6.1.1 *(a) Laminar flow (b) Turbulent flow.*

Experiments show that a measure of turbulence can be given by the Reynolds Number Re. The Reynolds Number is dimensionless and arises from a consistent grouping of quantities that have been shown, by experiment, to influence the type of observed flow. At low values of Re, viscosity forces dominate and the flow is laminar. At high Re, inertia forces dominate and the flow is influenced by vortices and other instabilities.

$$\mathrm{Re} = \frac{\rho v l}{\eta}$$

In this equation, ρ is the density of the fluid, v is the velocity of the fluid particles, η is the viscosity, and l is the characteristic length. The characteristic length depends on the physical situation. For a tube, it is most likely the diameter. For flow over a flat plate, it is the dimensions of the plate in the direction of flow.

For a tube of diameter equal to the length, $l = D$, Re < 2000 and the flow is probably laminar, When Re > 3000 the flow is probably turbulent.

The Reynolds Number is often combined with other dimensionless indicators (such as the Prandtl, Nusselt and Grashof numbers).

If the fluid is incompressible, then the mass rate of flow entering one end of the pipe is equal to the mass rate of flow out the other end. The mass of fluid Δm passing some point P during time interval Δt is:

$$\begin{aligned} \Delta m &= \rho V \\ &= \rho(Av\Delta t) \end{aligned}$$

Where v is the velocity of the fluid. Thus, for any two points P_1 and P_2 along the pipe,

$$\begin{aligned} \rho A_1 v_1 \Delta t &= \rho A_2 v_2 \Delta t \\ A_1 v_1 &= A_2 v_2 \end{aligned}$$

This is the equation of continuity.

The viscosity of a fluid is an indication of its resistance to flow.

When a fluid is held between two surfaces, and the bottom surface is fixed, and a force applied to the top surface to make it move, there is a change in velocity in the molecules comprising the fluid from the top surface to the bottom surface through the section of the fluid. Such a situation happens frequently in automotive engineering in sliding surfaces, such as piston rings sliding within the cylinder bore, a crankshaft rotating in the main bearings, steering ball joints in their housings, the meshing of gear teeth, etc.

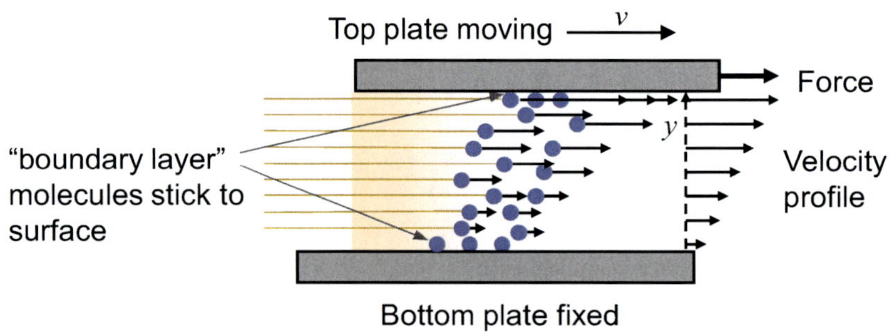

Fig. 6.1.2 *Velocity profile between two surfaces moving relative to each other.*

The rate of change of velocity with increasing distance through the fluid from one surface to the other is the velocity gradient and is usually a constant. A force needs to be continually acting to maintain velocity of top surface against viscous drag. This force, when spread out over a unit ross-sectional area, is an applied shear stress. Experiments show that there is often a linear relationship between the applied shear stress and the velocity gradient. The constant of proportionality is called the coefficient of viscosity η and has the units Pascal $\times$ seconds (Pas). Water has a viscosity of about 1 mPas, while a lubricating oil might have a viscosity of 5 mPas.

For the flow of a fluid in a narrow tube, the volume passing through the tube per second Q depends on:

- pressure difference Δp
- radius of tube (R)
- length of tube (l)
- viscosity of fluid η

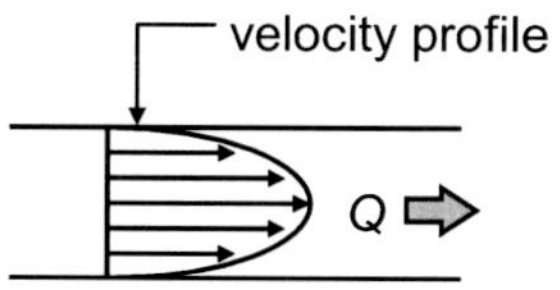

Fig. 6.1.3 *Velocity gradient through fluid flow Q in a tube of radius R.*

Assuming streamline flow, the relationship between the flow rate and these variables is called Poiseuille's law:

$$\dot{Q} = \frac{\pi \Delta p R^4}{8 \eta l}$$

The important feature of this equation is the dependence of flow rate on the 4th power of the radius.

A very important equation relates the pressure at any point within the pipe with the velocity of the fluid within it.

$$p_1 = \rho g h_1 + \frac{1}{2}\rho v_1^2 = p_2 = \rho g h_2 + \frac{1}{2}\rho v_2^2$$

This is the Bernoulli equation. The significance is that if the velocity of the fluid is made to pass through a restriction, the equation of continuity predicts an increase in velocity at the restriction. The restriction is most often called a venturi. Bernoulli's equation indicates that an increase in velocity in the venturi is accompanied by a decrease in pressure. It is this decrease in pressure that causes a carburetter to draw fuel through the main jet into the air stream. The velocity of the fuel through the main jet depends on the 4th power of its radius, and so, it can thus be appreciated that the profile of the needle of an SU carburetter alters significantly the rate of flow of fuel through the jet.

For an automotive fuel system, the nature of the flow of a gas through various pipes and orifices is of extreme importance. Air flows through an air cleaner, through a venturi, through a manifold, and exist via another manifold and a long pipe and muffler. The rate of flow is determined by the dimensions of the passages and the density and viscosity of the gas. In most cases, a high rate of flow for a given pressure difference is desired.

An inlet valve in a cylinder head is made larger than the exhaust valve because the pressure difference across it is the vacuum generated by the piston on the intake stroke – limited to something less than atmospheric pressure. On the other hand, the exhaust valve need not be so large since the pressure difference across it arises from the temperature of the exhaust gases and the rising piston on the exhaust stroke.

Carburetion, manifolding, and exhaust is all about gas flow. The dependence of flow on the diameter of the passage means that only slight changes are needed to make a substantial difference in the flow rates. The objective throughout is to design these components for maximum volumetric efficiency.

6.2 Combustion

In an internal combustion engine, hydrocarbon fuel and oxygen are ignited by the spark plug to provide pressure in the combustion chamber. The ratio of the mass of air to the mass of fuel is very important and is called the air-fuel ratio or "mixture". Theoretically, a little over 14.5 times the amount of air (by mass) than fuel (the actual ratio depends on the blend of the fuel) is needed for complete combustion. At this mixture ratio, all the hydrocarbons in the fuel combine with all the oxygen in the air during combustion and there is no free oxygen or hydrocarbons left over. The products of combustion under these conditions are carbon dioxide and water. In practice, this never happens, and the products of combustion are carbon dioxide (CO_2), water, carbon monoxide (CO), unburnt hydrocarbons (HC), and oxides of nitrogen (NO_x). The combination of unburnt hydrocarbons and oxides of nitrogen, with sunlight, produce photochemical smog. If the mixture is too weak, or too rich, the fuel will not burn at all and flat spots and rough running are the result.

At idle, (especially when the engine is cold), the mass of air and fuel (i.e. the "charge") coming into the combustion chamber is a minimum. Near the cylinder head walls, some of the liquid fuel doesn't completely get surrounded by oxygen. When the charge of the air-fuel mixture is greater (such as in wide open throttle) there is still some fuel near the walls of the combustion chamber, but this is a smaller proportion in comparison to the idle condition. Fuel at this location may not get completely burnt and instead of forming CO_2, will form CO instead, which can be measured in the tailpipe emissions by a CO meter. Some fuel may not burn at all in which case it passes out the exhaust system as free hydrocarbons (HC). The level of CO in the exhaust emissions is usually about 4% at idle for a pre-emission control car, dropping to about 1% at part throttle.

A major disadvantage of liquid hydrocarbon fuel is the requirement that liquid fuel be vapourised before it enters the combustion chamber. If liquid appears in the combustion chamber, not only will the hydrocarbon molecules inside the drop not be next to an oxygen molecule, ready for combustion at the command of the spark plug, but the spark plug will become wet and cease to function.

It takes heat to convert a liquid into a vapour, as one can readily experience by the cooling effect of perspiration. Thus, carburettors tend to run cold as heat is used to vaporise the fuel entering the inlet manifold. The original Mini engine was turned around 180 ° because when the carburettor was at the front, it got so cold that it iced up.

Often, car manufacturers connect the inlet and exhaust manifolds together so as to form a hot spot directly under the carburettor to help with vaporising the fuel. Others employ water heating of the manifold, particular in cross-flow head designs where the inlet manifold is on the other side of the cylinder head than the exhaust. The unfortunate side effect of this is that the density of charge of air/fuel entering the combustion chamber is reduced, thus reducing the available maximum power output.

The *mass* of air entering the carburettor, and hence into the combustion chamber, is dictated by the throttle plate angle. The mass of fuel being delivered by the carburettor depends on the diameter of a tube (the jet) through which the fuel passes from the float chamber into the air stream, and the pressure difference between one side of the jet and the other. The pressure on one side of the jet is atmospheric (the side that is connected to the float chamber) and the other side, is at the pressure at the venturi in the carburettor. The air passage of air past the venturi increases in velocity and decreases in pressure. It is the decrease in pressure (compared to the float chamber side of the jet) that causes liquid fuel to rise up through the jet into the air stream – although it is more accurate to say that the increased pressure on the float chamber side pushes the fuel through the jet. It is important to know that the pressure, or depression, at the throat of a venturi depends on the flow rate (in litres per minute) and the size of the venturi. The greater the flow rate, the lower the pressure compared to that on the upstream and downstream sides of the venturi. The narrower the size of the venturi, the lower the pressure.

At idle, a relatively small amount of fuel flow is required. At part throttle, or full throttle, a greater amount of fuel flow is required. Either the pressure difference, or the jet diameter can be increased to accomplish this. At acceleration, the throttle is opened quite rapidly. Air, having a lower density than liquid fuel, rushes through the carburettor and into the combustion chamber. The fuel, being a liquid, is at the same time being sucked through the jet in the carburettor but is slower to move than the air. The mixture in the combustion chamber is therefore weak (air-fuel ratio too high). To maintain a combustible mixture, the amount of fuel entering the air stream has to be temporarily increased.

6.3 The SU Carburetter

6.3.1 Introduction

The SU (Skinners Union) carburettor is a common fixture on BMC vehicles and indeed, Nuffield purchased the company in the mid 1920s. It is a simple device, but as with all simple devices, it is the second order terms that make it work in practice.

In an SU carburettor, the diameter of the jet and the size of the venturi are altered to let more or less fuel through it in response to air flow. The diameter of the jet is altered by putting a tapered needle down through the middle of it. The more the needle is inserted, the narrower the effective diameter of the jet. The needle position is determined by the height of a piston to which the needle is connected. The operation of this piston is important. The significance is that this piston can be held at nearly any position from fully down, or closed, to the fully up position with much the same amount of force. The force has to overcome the mass of the piston, and the relatively soft damper spring. The piston itself acts as a restriction to the air flow. That is, it is itself acting as a variable venturi, or narrowing, of the passage of air through the carburettor. This venturi is in fact a rectangle of constant width and variable height. The upper side of the piston is ported to the pressure at the venturi narrowing.

At a particular throttle opening, a certain amount of air in litres per minute is drawn through the carburettor. At idle, the piston is initially down. If the throttle is opened slightly, engine manifold vacuum tends to pull more air through the rectangular venturi formed by the piston and the throttle body. The greater flow rate of air through this venturi results in a *decrease* in pressure, which lifts the piston up – thus opening the venturi and tending to bring the pressure back to its former level.

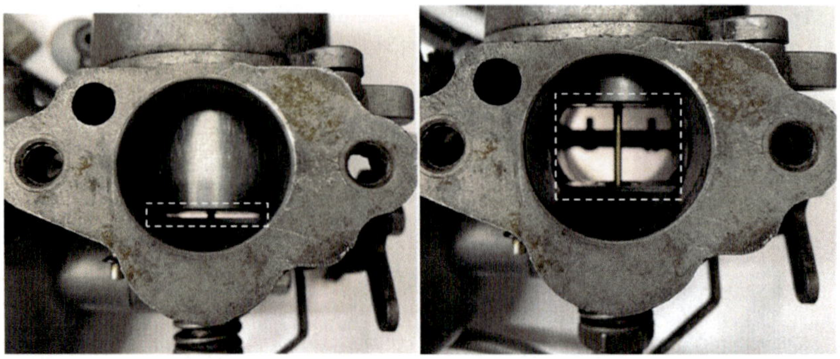

Fig. 6.3.1 *Full size venturi at wide open throttle, narrow venturi at idle.*

The piston will not drop, *because it takes the same amount of pressure to hold the piston at any position* so it just stays in the new position. Since the piston has now been raised, more air, in litres per minute, can flow through the larger venturi as required by the engine.

Since the piston has been raised, the needle, which is tapered, allows more fuel to be drawn from the float chamber, thus ensuring that the correct air-fuel mixture is maintained. The actual pressure at the venturi, and at the jet, is the same as before.

When the throttle is closed, the air flow through the carburettor decreases. This *increases* the pressure at the venturi and the piston is forced back downwards. In doing so, the height of the venturi decreases and tends to restore the pressure back to its former level. The piston then settles at a height appropriate to the air flow – the pressure at the venturi is back to its former level.

Since the pressure at the venturi and hence the jet, is constant, no matter what position the throttle is in, manufacturers can have some confidence in shaping the needle in the knowledge that they will also know what effective diameter of jet will provide a known quantity of fuel. SU carburettors fall into a class of what are called Constant Depression types, of which the CD Stromberg is a similar example, employing a diaphragm instead of a piston.

When the engine is cold, in a conventional carburetter, proper air fuel mixture is maintained by the choke. The choke is a flap valve that restricts the passage of air through the carburetter and puts more direct manifold vacuum on the jet. In the SU carburettor there is no choke as such although it is still labelled that way on the control cable. The choke cable pulls the jet down from its normal position a few mm which, due to the taper of the needle, makes the effective jet diameter larger.

During acceleration, the mixture is made richer by damping the piston so that more suction is temporarily applied to the jet which increases the fuel flow through it. When the "acceleration" is finished, the piston settles down again to a new position which depends on the throttle plate position as described previously.

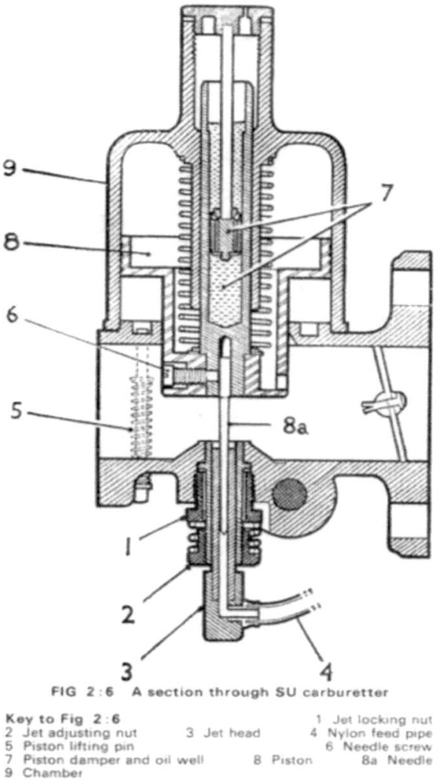

Fig. 6.3.2 *Carburetter cross section.*

6.3.2 Dashpot

There's an interesting issue in relation to the oil level in the SU dashpot that deserves comment. BMC workshop manuals usually state *"pour in sufficient quantity of engine oil to bring the level to about ½ in. (13 mm) above the top of the hollow piston rod."* Intereurope workshop manuals repeat this advice verbatim. Scientific Publications manuals offer *"pour in about a teaspoonful of light engine oil"*, while Haynes recommends to *"top up the damper"*. Autoexpress provides a helpful *"top-up with engine oil to a level ½ inch (13 mm) above the top of the hollow piston rod"* but then adds *"on dust-proofed carburettors, with a transverse hole across the neck of the chamber and no vent hole in the damper cap, the level must be ½ inch (13 mm) below"*. Interestingly, all those publications show a cutaway view of the device which clearly show the oil level as being *below* the top of the piston rod. In an early Mini owners handbook, 1962, the instructions are *"pour oil into the hollow piston rod until the level is ½ in (13 mm) from the top of the rod"* and the accompanying picture clearly shows the level being below the top of the rod. If oil is poured ½ above the top of the rod, it will most assuredly spill out the breather hole when the piston damper rises, and any excess will soon be sucked down into the piston chamber where it will eventually be consumed by the engine. The only possible advantage of over filling the piston rod is to provide some lubrication of the piston rod from time to time during normal routine service.

CHAPTER 6. FUEL SYSTEM 127

The nature of the oil to use is also open to some interpretation. "Light engine oil", "engine oil" and "thin engine oil" all mentioned in various BMC instructions. Some practitioners promote the use of automatic transmission fluid. BMC cars were designed when oils were not the multi-grade products they are today. When the engine is cold, it would be helpful to have the oil a little thicker (higher viscosity) to provide a little more rich a mixture for acceleration compared to when the engine is at operating temperature – where the damper oil can be thinner. Perhaps a single viscosity oil, such like Singer sewing machine oil would be a good substitute and is widely used for this purpose. Various BMC documentation indicates SAE 20 oil for the carburettor dashpot, somewhat lighter than what is used in the engine.

6.3.3 Needle

As discussed previously, the flow of liquid through a tube depends on the radius of the tube raised to the fourth power, everything else being equal (i.e. for the same pressure difference). This means that the slightest wear on either the jet or the needle is magnified greatly in terms of the effect it has on mixture. This is why it is necessary to have the needle centered in the jet – to prevent wear. It is also important that the correct needle, jet and piston spring be fitted. The taper of the needle is critical to performance and the practice of running sandpaper over the needle in an attempt to extract more performance should be avoided.

For the HS2 carburetter fitted to the Mini range, needles are available in several grades. Whether one needle is considered rich or lean depends on the vehicle model, but generally needle EB is considered standard in Australia.

Running condition	Morris 850	998cc	1098cc	1275 TWIN
Rich	M	M	H6	AH2
Standard	EB	GX	AN	M
Lean	GG	GG	EB	EB
Damper spring	Red	Red	Red	Blue

Table 6.3.1 *HS2 needle specifications single carburetter (Morris 850).*

Needles are specified in terms of their diameter at every 1/9" steps down from the thickest cross section at the shoulder tapering to the thinnest.

Step	Diameter (ins)				
	EB	M	GG	H6	AH2
1	0.089	0.089	0.089	0.089	0.089
2	0.0855	0.0855	0.087	0.0855	0.085
3	0.0835	0.0832	0.0845	0.082	0.082
4	0.0815	0.0808	0.0825	0.080	0.0794
5	0.0795	0.0785	0.0808	0.078	0.077
6	0.0777	0.0763	0.079	0.076	0.0748
7	0.0762	0.0745	0.078	0.074	0.0726
8	0.075	0.0726	0.077	0.072	0.0704
9	0.074	0.0707	0.0758	0.070	0.0683
10	0.073	0.0688	0.0745	0.068	0.0662
11	0.072	0.0669	0.0732	0.066	0.064
12	0.071	0.065	0.072	0.064	0.062
13					0.060

Table 6.3.2 *Needle measurements, 0.090" jet diameter.*

6.4 Detonation

One of the most limiting factors that affect engine design is the onset of detonation, "pinking" and "pinging".

Of course ignition timing and carburetion work together to provide that smooth ride we are all accustomed to in our BMC vehicles. A common cause of damage and problems is pre-ignition, detonation, "pinking" and "pinging". All these terms *almost* mean the same thing. Although pinging and pinking might describe the sound, the processes of pre-ignition and detonation are not the same.

A very high compression ratio may often result in the expanding gases compressing the unburnt mixture ahead of it to the extent that portion of the unburnt charge may spontaneously ignite. If there is a hot spot in the combustion chamber, from say a glowing deposit or corner, then this may ignite the mixture before the spark plug fires. In both cases, there are two flame fronts and when they meet, the resulting turbulence creates shockwaves of high magnitude which overload the mechanical parts of the engine and cause damage. Running-on after you turn off the ignition on a hot day is a sure sign something needs attention in the cylinder head or cooling system. If you look carefully, the engine often runs backwards when running on which may cause problems in the valve train. In some cases running on can just be a result of the idle speed setting being too high. Some later model emission control cars have a throttle stop solenoid to close of the throttle completely when the ignition is turned off to avoid this undesirable event.

In contrast to the complexity of a modern fuel injection system, the simplicity and reliability of an SU carburetter represents a significant part of that pleasurable BMC ownership experience that is denied to owners of more modern vehicles.

CHAPTER 7. SUSPENSION

7.1 Suspension Geometry

7.1.1 The Reason for a Suspension System

We all tend to take it for granted that motor vehicles have a suspension system but tend to forget why this might be so. The simple answer is that a suspension system provides a comfortable ride. While this might seem trite, the concept is of extreme importance and, together with handling and steering characteristics, makes up a considerable portion of motor vehicle design.

Research shows that in relation to vibrations transmitted to the passengers, the higher the frequency, the smaller the amplitude of the vibrations are tolerable to maintain comfort. On the other hand, for low frequency vibrations, passengers can cope with larger amplitudes. There is an empirically derived "comfort curve" that provides limits for designers to work with when designing spring rates and damping characteristics.

The suspension system essentially deals with the movement of the body in three directions as a result of forces acting on the centre of mass through cornering, braking, and variations in the road surface. Another important factor is the effect of the variation in static weight of the vehicle as passengers and goods are added or removed.

In this Section we will not discuss commonly studied topics such as castor, camber and king pin inclination, but examine some less-known concepts, in particular, roll centres and mass.

7.1.2 Roll Centres

One of the most significant design parameters of a motor vehicle suspension system is the disposition of the front and rear roll centres. A line which joins these two points is called the roll axis and is the axis about which the vehicle body rolls when cornering. Such is the influence this axis has on handling that it behoves us to learn more about it.

Consider the simplified geometry shown in Fig. 7.1.1. At rest, the weight W of the car acts through the centre of mass and is supported by springs, and then the tyre, with the reaction forces $W/2$ shown on each side of the vehicle (ignoring the weight of the wheels and axle and the other wheels). Suppose now a sideways force F is applied at the centre of mass. Fig. 7.1.1 (b) shows the reaction forces at the point of attachment of the springs. Since these reaction forces act at a distance h from the centre of mass, there is a turning moment of magnitude Fh created.

To maintain equilibrium, it is not sufficient to just balance the sideways forces, but the turning moment must also be balanced. This is done by an increase in the right hand vertical force (arising from compression of the spring on that side) and a decrease in the left hand vertical force (arising from a relaxation of the spring on that side), both of these acting through a distance *t*. That is, the body tilts around a point RC, which is the roll centre.

Thus, the spring forces change by an amount *q*, which is added to (right hand side) and subtracted from (left hand side) of the upwards spring forces. The magnitude of *q* depends on the magnitude of the sideways for *F* and the moment arm *t*. A moment's thought will show it doesn't matter where the moments are referred to, the resulting magnitude of *q* will be *Fh*/*t*. Since the upward deflection of the spring at a will be matched by the downward deflection of the spring, then the centre of rotation must be midway between.

By a similar argument to the above, where the tyres take the place of the springs, the reaction forces at the road change by an amount *p* = *FH*/*T* where *T* is the track, and *H* is the height of the centre of mass above the ground and the centre of mass rotates around the centreline at ground level. Evidently, the reaction forces at the road depend not only on the sideways cornering forces, but also the distance between the centre of mass and the roll centre.

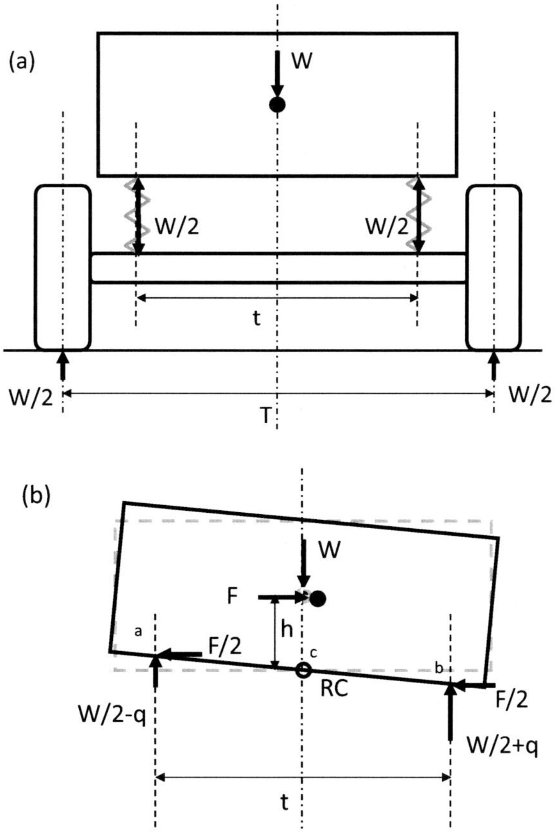

Fig. 7.1.1 *Location of front roll centre for a simple suspension system.*

Most vehicles do not have a suspension geometry of this type but employ more complicated arrangements of links as shown in Fig. 7.1.2. The upper and lower links may be parallel, or they might converge at a point inside or outside the wheel. Fig. 7.1.2 shows the position of the roll centre for a variety of suspension geometries. Note that for parallel links that are parallel to the ground, the roll centre is at ground level even if the links themselves are above.

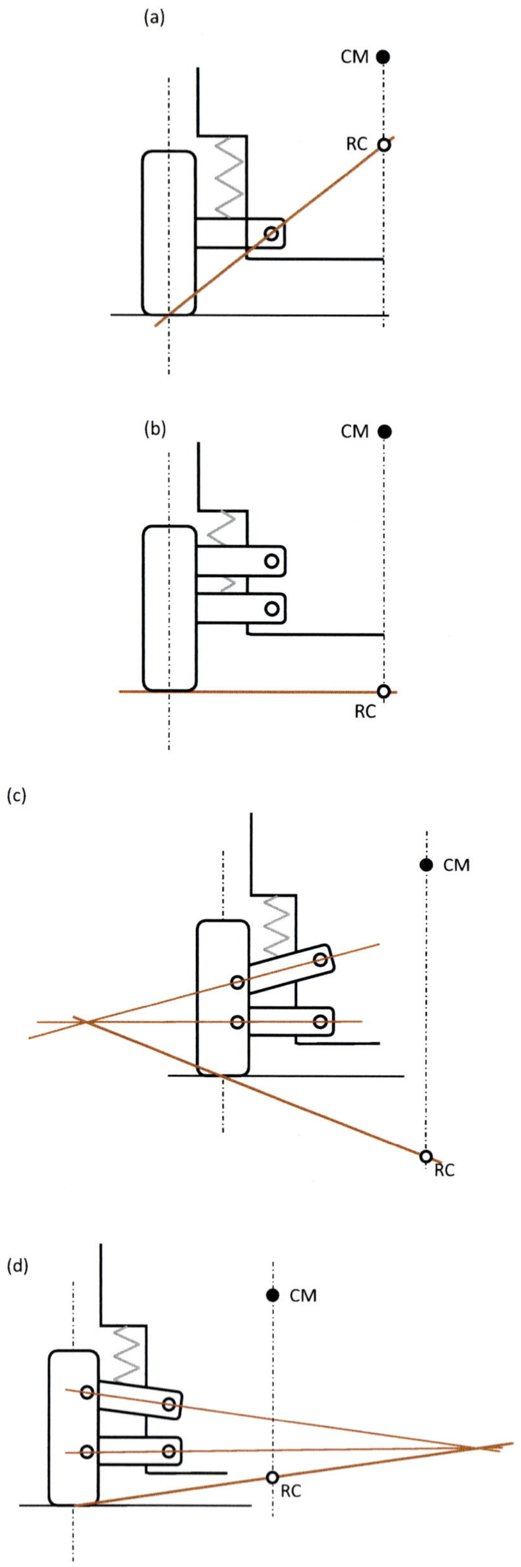

Fig. 7.1.2 *Location of front roll centre for various suspension geometries.*

As can be seen from Fig. 7.1.2, the roll centre may be above, at, or below ground level. When the vehicle rolls, the centre of mass shifts position, and the geometry of the linkages also change, and the roll centre thus also moves.

What then are the consequence of the roll centre location?

A low roll centre results in large roll angles. In this case, spring deflection may be excessive compared to suspension travel available and may result in bottoming or topping out against the bump stops. A low roll centre also results in a shift in the position of the centre of mass such that the reduction in force on the left hand side of Fig. 7.1.1 may reduce to zero (lifting of the inside front wheel, and so adversely affecting handling leading to understeer.

A high roll centre reduces the roll angle, but the wheels move sideways with consequent slip at the contact patch which adversely affects cornering.

Manufacturers usually design for a fairly low roll centre at the front and a higher one at the rear.

In a Morris Mini, the upper and lower arms are essentially parallel as shown in Fig. 7.1.3. The roll centre is found at the intersection of a line drawn parallel to the links, beginning at the centre of contact with the road, to the vehicle centreline. In this case, it can be seen that the roll centre is fairly low in relation to the centre of mass of the vehicle. The rear roll centre is drawn through the axis of the stub axles. As is usual custom, the front roll centre is lower than the rear roll centre.

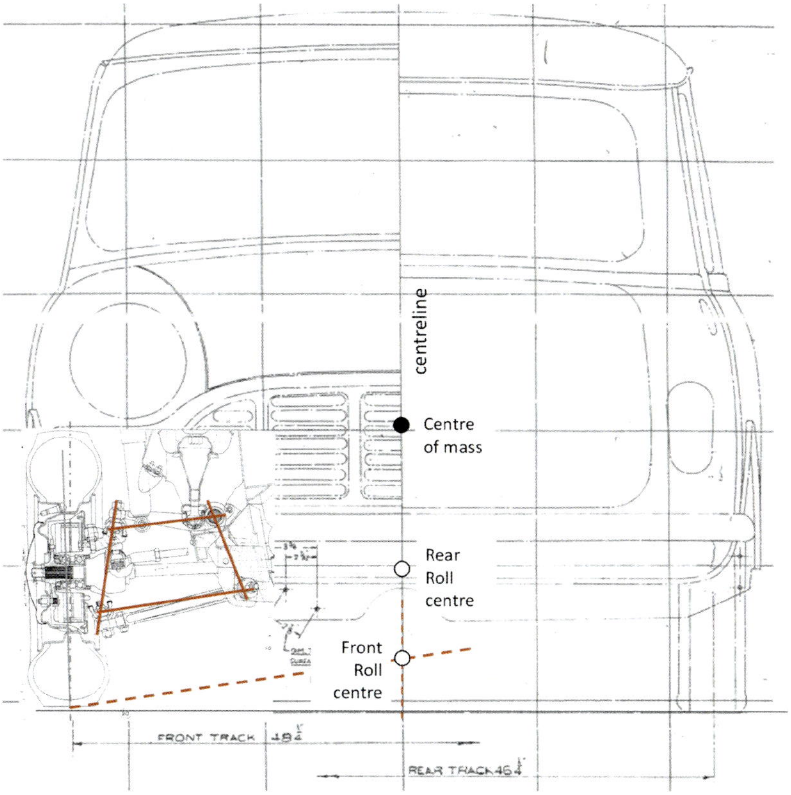

Fig. 7.1.3 *Front suspension detail, Morris Mini.*

The roll axis therefore looks like that shown in Fig. 7.1.4.

Fig. 7.1.4 *Roll axis Morris Mini.*

For the Austin 1800, the front roll centre is indicated by Fig. 7.1.5 with construction lines shown for the front roll centre determination.

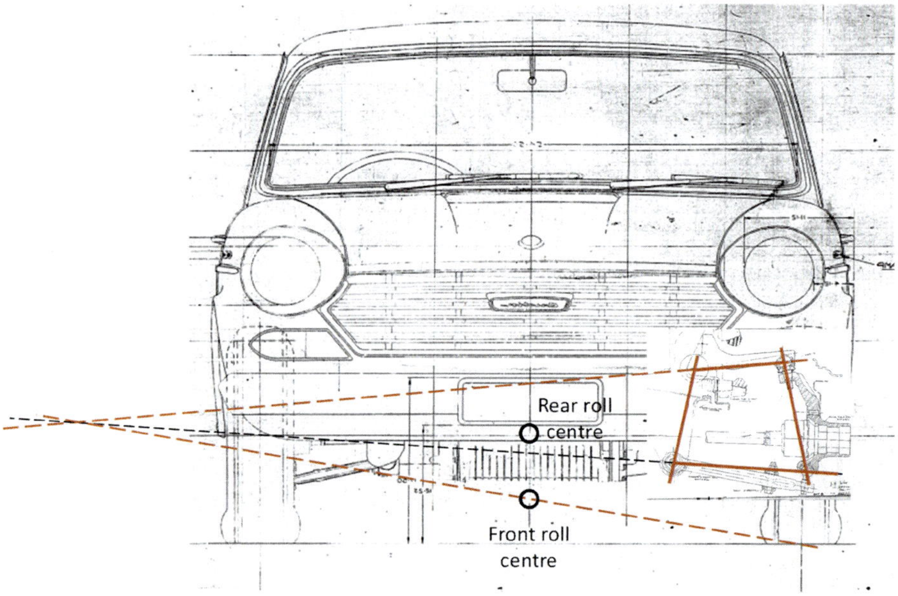

Fig. 7.1.5 *Front roll centre, ADO17.*

We might ask ourselves how is the point of the centre of mass determined? The lengthwise position is easily found by simply weighing the vehicle front and rear and determining the weight distribution. For a Morris Mini, the weight distribution front to rear is about 60:40 and it is therefore no surprise to find that the jacking tube in the sills is positioned at this position. Determination of the height of the centre of mass above the ground is not so easily determined. One method is by tilting the vehicle and finding the point of balance before tipping over. Some error will be introduced as a result of the deflection of the suspension and tyres, but it is a simple, low cost, measurement. Factory data shows that the centre of mass height for a Morris Mini is 510mm.

As far as comfort is concerned, it is the relationship between the sprung and unsprung mass that is important, and we should therefore examine the issue of vehicle mass attentively.

7.1.3 Vehicle Mass[18]

7.1.3.1 Importance of Vehicle Mass

Vehicle mass is probably the single most important aspect of vehicle design. All things considered, the kerb weight of a vehicle determines its minimum selling price. Vehicles are produced based on market decisions, and as such, the vehicle mass is thus determined. In calculating vehicle mass, Table 7.1.1 shows the distribution for a Morris 1100, ADO16, based upon a classification introduced by General Motors.

Component Group	% of total mass
Body complete	43.8
Front Suspension	4.46
Rear Suspension	4.46
Steering	1.32
Cooling System	0.8
Braking System	4.46
Fuel and Exhaust Systems	1.78
Road Wheels	7.75
Heating System	0.7
Electrical and Instruments	3.05
Power Unit Complete	23.0
Fuel and Water	4.08

Table 7.1.1 *Distribution of vehicle mass, ADO16.*

Mass savings affect fuel consumption, vehicle performance and parasitic mass. At BMC, it is shown that fuel consumption is related to vehicle mass according to:

$$L/100km \sim Mass^{0.4} + Eng.Cap.^{0.25}$$

Acceleration time can be shown to be proportional to the square root of the mass and the engine capacity:

$$Acceleration\ time \sim \sqrt{Mass} \times \sqrt{Eng.Cap.}$$

Parasitic mass is that extra mass needed as a result of an increase in mass of some other component. For example, a larger engine will require heavier front suspension, wheels, engine mounts, radiator, fuel tank, etc. Generally speaking, any increase in mass of one component requires a 50% penalty for parasitic mass.

7.3.1.2 Unsprung Mass

As far as the suspension system is concerned, vehicle mass falls into two main classes. That which is supported by the springs, the sprung mass, and that which is not, the unsprung mass.

Unsprung mass includes wheels, tyres, brakes, and a proportion of the suspension springs, arms and linkages. Sprung mass is the body including seats and trim, glass, fuel, water, etc.

The mass of the vehicle has to contend with various loads, which include: static load, lateral loads (cornering), rut load, bump load, braking load, rebound load, damper load, and engine mounting load. Some of these can be of the order of 4 – 6g.

---

[18] Acknowledgement: Ray Habgood, Barry Anderson and Graham Hardy (dec)

Bearing in mind that the purpose of the suspension is to provide a comfortable ride for the passengers (who form part of the sprung mass), it should be appreciated that the unsprung mass is called upon to follow any irregularities in the road surface. Thus, the unsprung mass suffers deflections of varying frequencies and amplitudes. It is desirable that the sprung mass be isolated from these movements by virtue of compression on elongation of the springs.

But, masses supported by springs bring with them various complications, the most significant being a tendency to oscillate by themselves at their natural, or resonant, frequency. As well, the movement of the unsprung masses, often being quite substantial, involve significant inertia forces which have to be accommodated as well. It is desirable therefore that the ratio of the sprung mass to the unsprung mass be made as large as possible and the spring as soft as possible. For this reason, modern tendency is for motor vehicles to have lightweight alloy wheels in contrast to earlier steel wheels. Some expensive models feature inboard brakes as well. All designed to reduce unsprung mass.

7.3.1.3 Vehicle Weight

Vehicle mass or "weight" is not a term so easily pinned down. The following table lists the different design parameters used in motor vehicle design.

Weight type	Definition
Kerb weight	Vehicle full equipped with oil, water, tools, spare wheel and full fuel tank. No passengers.
Dry weight	As with kerb weight but no fuel or water.
Light running weight	Vehicle fully equipped with 2 gallons of fuel, 2 front seat occupants (150lb each), and 50lb luggage placed centrally in the boot.
Normal weight	Vehicle fully equipped with full fuel tank, 2 front seat occupants and 2 rear seat occupants, and 100lb luggage in the boot
Maximum weight	Vehicle fully equipped with three front seat occupants (for bench seat) or two front seat occupants (bucket seats) and 3 rear seat occupants and 150lb in the boot (bench seat) or 125lb luggage in the boot (bucket seats).

Table 7.1.2 *Vehicle weight.*

Some examples for BMC/Leyland vehicles are s shown in Table 7.1.3.

Vehicle		Dry Weight (kg)	Kerb Weight (kg)
YDO21	Leyland Mini	634	656
YDO21V	Leyland Mini Van	606	627
YDO22	Leyland Mini S	656	677
YDO18	Leyland Moke	602	625
YDO24	Leyland Marina 4D/S/Man	898	943
YDO24	Leyland Marina 4D/S/Auto	913	958
YDO24	Leyland Marina 2D/S/Man	884	929
YDO24	Leyland Marina 2D/TC/Man	895	940
YDO19	Austin Tasman Man	1098	1137
YDO19	Austin Tasman Auto	1124	1162
YDO19	Austin Kimberley Man	1123	1161
YDO19	Austin Kimberley Auto	1143	1181

Table 7.1.3 *Dry weight and kerb weight for Leyland passenger vehicles 1973*[19].

[19] Habgood, 1973

The weight distribution between front and rear wheels plays an important part in the design of the suspension since it alters the handling characteristics of the vehicle with respect to oversteer and understeer during cornering. The distribution may change according to loading conditions and sample values for YDO25, Leyland Marina, are shown in Tables 7.1.4, 7.1.5 and 7.1.6.[19]

	2D	4D
4 Cyl	55.5	54.2
6 Cyl	57.2	55.9

Table 7.1.4 *% front weight distribution YDO25 Marina.*

	Weight to be added (kg)		
Weight condition	**Total**	**Front**	**Rear**
Dry weight	-44	+1	-45
Registration weight	-32	+6	-38
Light running weight	+127	+53	+47
Design weight (3 occupants)	+238	+88	+151
Maximum weight (4 occupants)	+318	+101	+217
Maximum weight (5 occupants)	+397	+114	+283

Table 7.1.5 *Weight conditions, taking kerb weight as the baseline, YDO25.*

	Weight to be added (kg)		
Item	**Total**	**Front**	**Rear**
Petrol	38	7	40
Water	5	6	0
1 front seat occupant	68	39	29
1 rear seat occupant	68	15	53
Luggage in boot (11kg)	11	2	13

Table 7.1.6 *Additional weights taking kerb weight as a baseline, YDO25.*

7.1.4 Suspension Springs

The nature of the action of springs dates back to Robert Hooke (1605-1703) who uttered the famous words "ut tensio, sic uis" and has become familiarly known as Hooke's Law. This is a law of wide applicability and governs the deformation of solids under load, from buildings to motor cars as well as springs. The meaning is that if you have a spring, and you compress it with a force F, then it will deflect a distance x. If you double the force to $2F$, the spring will compress twice the amount, $2x$. Although this might sound obvious today, at the time, no one had any idea how solids behaved when placed under load and by careful experimentation with springs and weights, Hooke figured it out.

For us BMC front wheel drive vehicle owners, the significance is that Hooke's law does NOT apply to the rubber springs. The rubber spring is "non linear". What this means is that if you double the force on the spring unit, the amount of deflection is less than $2x$ – as if the basic "stiffness" of the spring increases the more you load it up. This is precisely why Issigonis went for rubber springing in the Mini – because for such a small car, the addition of four occupants adds very significantly to the weight of the whole affair and ordinary springs would just deflect too much. Making the springs stiffer would help, but then ride quality would suffer. So, he put in rubber springs which are "softer" when unladen compared to becoming "stiffer" when laden.

7.1.4.1 Simple Harmonic Motion

If a mass is mounted on a spring and set into oscillation, the force applied to the mass varies in time according to:

$$F = ma = -m\omega^2 y$$

where the displacement of the body y is a sine wave

$$y = A\,sin(\omega t)$$

However, the product $m\omega^2$ is actually a constant "*k*". Thus:

$$F = -ky$$

The minus sign indicates that the force acting on the mass is in a direction opposite to the displacement of the mass and acts so as to bring the mass back to the equilibrium position. The magnitude of this restoring force is a function of displacement of the mass from the equilibrium position. *k* is often called the spring or force constant.

The combination of periodic motion and a restoring force whose magnitude depends on the displacement from the equilibrium position is called simple harmonic motion.

Now,

$$\omega = \sqrt{\frac{k}{m}} = 2\pi f = \frac{2\pi}{T}$$

Where *f* is the frequency in Hz and *T* is the period of the motion.

Since,

$$T = \frac{2\pi}{\omega}$$

then

$$T = 2\pi\sqrt{\frac{m}{k}}$$

Note that the frequency and the period of oscillation only depends on the mass and the spring constant and not on the amplitude. The frequency ω is the resonant frequency of the system.

These equations tell us that as the mass of the vehicle increases (such as by taking on additional passengers), then the natural frequency of the system decreases. In practice, this means that the amplitude of the oscillation of the sprung mass is equal to that of the road irregularities at low frequencies, then increases up to some maximum at the resonant frequency, and then decreases after that. For example, if we have a road in which the height of the irregularities is constant, but their spacing is getting shorter, then the vertical displacement of the spring mass will increase to some maximum, then decrease as shown in Fig. 7.1.6.

Since the objective is to reduce the vertical motion of the body, this is achieved by either having a large mass or a small spring constant (i.e. a soft spring).

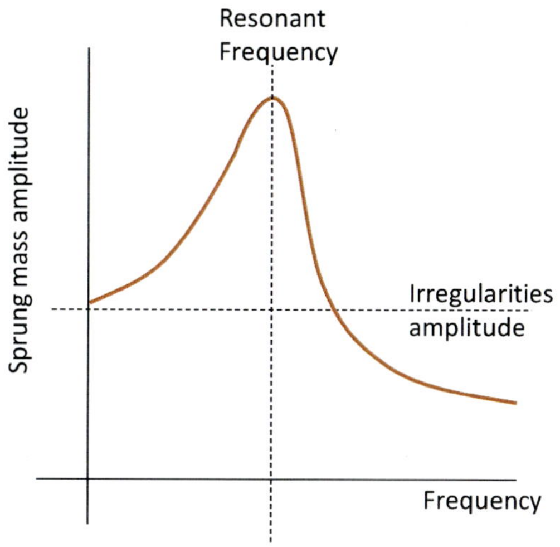

Fig. 7.1.6 *Vertical motion of vehicle body as a function of frequency of road irregularities.*

Of particular relevance to small cars is that the mass of the vehicle increases markedly in proportion to the addition of extra payload. This means that with four people on board, the lower the natural frequency, (and consequently, the sharper the "bandwidth"). In the Morris front wheel drive range, rubber springing is used because not only are the spring units small, but their effective spring rate increases as the mass they support increases. The spring rate is non-linear. It should be also appreciated that the deflection of the spring is not necessarily equal to the deflection of the road wheel due to the leverage effect of the suspension arms.

The spring stiffness is usually dictated by other things such as the available suspension travel, the likely payload to be carried, and the amount of body roll and pitch that is acceptable. The spring rate must then be made high enough to cater for suspension movement while at the same time providing an acceptable ride.

7.1.4.2 Dampers

In practice, the problems that would arise from the resonance condition are avoided by the fitment of dampers, or shock absorbers. The term damper is technically correct since from the point of view of the deflections involved, it is the spring that absorbs road shocks and that energy stored in the spring is then released into the damper.

Dampers absorb the energy of an oscillating spring converting the energy of vibration into heat. Rubber has an inherent damping function due to hysteresis in the molecular structure. Leaf springs also have a damping characteristic arising from friction between the leaves, but the effectiveness of these mechanisms is usually insufficient for automotive suspension use. Thus, the most common type of damper is a hydraulic device in which oil is forced to pass at high speed through small orifices. The viscosity of the oil causes it to absorb energy as heat. At low speeds, the damping effect is low, but increases as the square of the speed so that at high frequency oscillations of the suspension spring, the motion is more effectively damped.

Further, it is desirable that there be less damping on the compression stroke of the spring (so that the spring performs its function to accommodate road irregularities) compared to on the rebound.

7.1.4.3 Spung and Unsprung Mass

Now, imagine that the suspension is arranged so that the springs support the seats that we sit on, and all the rest of the vehicle body is rigidly connected to the wheels. Surely, you would say, we would enjoy splendid isolation from the road while the body around us bolts and crashes its way through and over any obstacle. But no. If you were to drive such a car, you would find that you would generally suffer great discomfort. The reason is that in this example, it is only the mass of the seats and occupants that is "sprung".

Now, when the road wheel rolls over a bump, the unsprung mass has to follow the wheel up. There is no other place for it to go. That is, this great mass is accelerated upwards. Sitting on our sprung seats, there is not much resistance or inertia on our side of the spring for the upwards motion to work against and so the spring only compresses a little, and we shoot up in the air along with everything else.

For maximum comfort, it is desirable that the unsprung weight be a minimum. When this happens, there is a great mass up on top of the spring which tends (due to its inertia) to stay where it is, and the energy coming from the upwards motion of the unsprung weight is absorbed by compression of the spring instead of going into raising a great mass upwards.

Not only are light alloy wheels providing us with lower unsprung mass, but reducing vehicle mass overall reduces other parasitic mass and improves vehicle performance. There is consequently a lot to be said for the use of "mag" wheels on any vehicle.

7.1.5 Design Parameters

An engineering analysis of the motor vehicle chassis involves suspension design parameters such as those shown in Table 7.1.7 (Leyland P76)[20].

Item	Front	Rear
Travel (Laden to M-M bump)	4"	3.62"
Travel (Laden to M-M Rebound)	4.5"	5.06"
Lock angle (inner wheel)	92.3 lb/in	102.8 lb/in
Lock angle (outer wheel)	65 cpm	70 cpm
Toe-out 20° lock	40°34°30'	
Turning circle	35'8"	
Camber	0°	
Camber (M-M Bump)	0° - 14' Pos.	
Camber (M-M Rebound)	3° - 30' Pos.	
Caster	1°	
King Pin Inclination	9° - 20'	
King pin offset at stub axle	3.97"	
King pin offset at ground	2"	

Table 7.1.7 *Suspension design parameters, design load of 4 persons plus 45kg luggage.*

[20] Habgood, 1972

7.2 Hydrolastic Displacers

7.2.1 Introduction

One of the more unique features of BMC vehicles of the mid-sixties is the hydrolastic suspension system developed by Alex Moulton.

This type of suspension was introduced in Australia in 1964 for YDO16 Morris 1100 and was carried over to the Mini range in 1965, and also the Austin 1800/Tasman/Kimberley. It was phased out in 1973 with the introduction of the Leyland Mini. In Australia, the hydrolastic units were manufactured by Dunlop.

It should be realised that although the movement of the suspension causes fluid to be displaced within the unit from the lower chamber to the upper chamber, and vice versa, the actual springing medium is the rubber cone insert just as in the case of the dry suspension fitted to ADO15.

The damping and rubber characteristics are tuned for each application. Identification of particular specification is usually made by painted bands on the outside of the unit which are easily seen (when new), but somewhat difficult to find in more modern times. Part numbers stamped into the casing can also be used in some cases to identify specifications. The coloured bands may be either near the top outside edge, or near the circumferential swage towards the middle.

In this Section, the distance *b* is that measured from the edge of the mould line of the rubber at the top inside of the unit to the centre point of the blend radius *d* with the rubber spring (in the unpressurised condition) as shown in Fig. 7.2.1. Although internally there are different specifications for the flap valve bleed diameters, the units cannot be dismantled for inspection and the distance *d* and the coloured bands are probably the only external means of identification available.

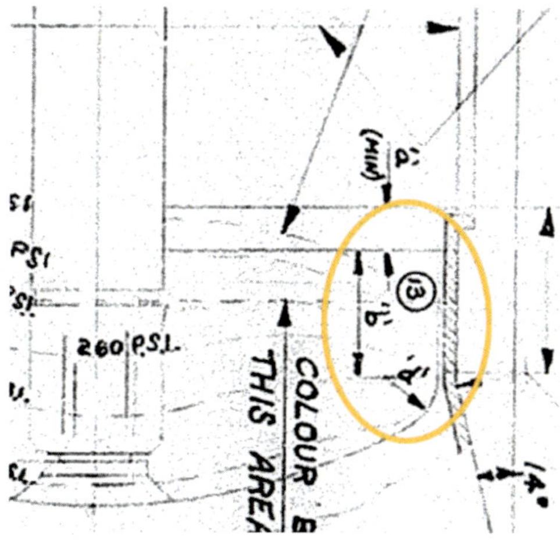

Fig. 7.2.1 *Hydrolastic unit detail showing distance b, 21A1477.*

Drawings show that Australian-made hydrolastic units have a different shape to the upper spring surface to those with UK part numbers, the former having a pronounced raised portion near the hose attachment.

7.2.2 ADO16 Morris 1100

For both front and rear suspension, hydrolastic unit AYG4076A is specified for YDO16. This is identified by a green spot. It has a nominal 6" diameter. The same displacer (but with different hose connection) is used for the rear of Austin 1800 models.

7.2.3 YDO5/YDO22 Morris Mini

YDO5 uses the same displacer unit front and rear, and is specified as 21A1703 (no band), 21A1477 (no band) and then 21A1804 (AYA4090) (from YMA2S2 16758) with single orange (white) band, all with distance *b* = 0.77" for both front and rear displacers. The displacer has an outside diameter of 4.84".

Part No.	Identification Bands	Distance "b" (inches)	Piston length A (inches)	Rubber Spec	Damping/ Bleed dia	Change Point
	Front and Rear					
21A1703	No band	0.77	1.785	NR	K420 1/8	
21A1477	No band	0.77	1.785	NR	K420 1/8	
21A1804 (AYA4090)	Single orange or single white	0.77	1.620	NR	K420 1/8	From YMA2S2 16758 to 30543
21A2008	Single green	0.77	1.620	NR	K420 1/8	30544 onwards

Table 7.2.1 *YDO5 displacer details, 21A1477.*

7.2.4 YDO6/YDO23 Morris Cooper S, Morris Mini Clubman GT

Several hydrolastic spring units are specified for the front suspension of YDO6, those being 21A1705 (single yellow band), 21A1811 (twin orange band), 21A1872 (single blue band), 21A2012 (single silver band), and 21A2010 (twin green bands) as well as AYG4113 (single brown band). All differ from that used in YDO5. The outside diameter is 4.84".

Part No.	Identification Bands	Distance (b) (inches)	Change Point
Front			
21A1705	Single yellow	0.77	To YKG2S2 1219
21A1811	Twin orange	0.77	YKG2S2 1220 – YKG2S2 1917
21A1872	Single blue	0.47	YKG2S2 1918 - YKG2S2 3716
21A2010	Twin green	0.77	YKG2S2 1918 - 3716
21A2012 or AYG4113	Single silver; **or single brown;** or single orange with brown dot	0.47	YKG2S2 1317 on
Rear			
21A1705	Single yellow	0.77	To YKG2S2 1219
21A1811	Twin orange	0.77	YKG2S2 1220 – YKG2S2 1917
21A1874	Twin blue	0.41	YKG2S2 1918 - YKG2S2 3716
21A2014 or AYG7061	Twin silver; or twin brown; or twin orange with brown dots.	0.41	YKG2S2 1317 on

Table 7.2.2 *YDO6 displacer details, 21A1477.*

7.2.5 ADO17/YDO13/YDO19 Austin 1800, Austin Tasman/Kimberley

The displacer for the rear suspension for the Austin 1800 range differs from that of the front for the saloon models. The 6" rear displacer is AYG4076A (same as Morris 1100) up to MKII models YAHS5 8944 and AYH7195 (identified with a yellow dot) thereafter. These are equivalent to UK parts 47H9543. For Tasman/Kimberley, the rear displacer unit AYB7107 is specified. This is similar to ADO17 but with a different bump/rebound characteristic and is identified by an orange spot.

For the front suspension, 6 ½" displacer unit 11H1128 is specified for Austin 1800 MKI models, and 11H1887 for MKII models (same as AYH4194 as used for Tasman/Kimberley), either of which can be replaced by 11H2158. The same displacer is used on the rear suspension for Austin 1800 utility models. For the Tasman/Kimberley range, the front 6 ½" displacer units are AYH4194. Displacers are identified by an orange spot.

7.2.6 Hose Data

A word should be made about the hydrolastic hoses. Hoses can be replaced, but it is important to do so with a high pressure hydraulic fitting, not crimped dog-eared clips or screwed clamps. The hoses are of single braid type, and it is important not to replace hoses with those overly stiff. The hoses are required to bend in an arc when fitted and if overly stiff hoses are used, they are likely to put undue strain on the fittings at either end or may chafe on other surfaces.

For the Mini range, the hose connection has 5/8 x 18 UNF thread for connection to the interconnecting bundy pipes. The hose has OD 0.680-0.710" and ID 0.365 – 0.380". The recommended minimum working bend radius is 2.5" and minimum burst pressure of 2000 psi.

For ADO16 and ADO17, the drawing indicates a hose connection with ¾ x 16 UNF thread. The recommended minimum working bend radius is 2.5". The hose is rated with a minimum burst pressure of 2000 psi, OD 0.870 – 0.900" and ID 0.5209 – 0.535".

7.3 The Hydrolastic System

7.3.1 Hydrolastic Fluid

Officially, there are four grades of hydrolastic fluid and the characteristics of these are:

1. A water based fluid for use in frost-free conditions in tropical territories where sanction has been granted in writing by the factory.
2. A low alcohol/water base fluid for use in light frost conditions in mainland Australia and in territories where sanction has been granted in writing.
3. An alcohol/water base fluid for use in severe frost conditions (restricted use in some export vehicles).
4. An alcohol/water base fluid of controlled viscosity developed as a special purpose fluid. The use of this fluid is subject to approval from B.M.C. Product Engineering Department.

Under no circumstances should any mineral based oil product be used. Brake fluid, engine oil, auto transmission oil, hydraulic jack oil, or even silicon fluid should never be used. If no proper hydrolastic fluid is available, water is acceptable with some anti-freeze added if temperature is likely to be low.[21]

Hydrolastic fluid has a specified composition. The specifications for each grade are given in Table 7.3.1:

	Grade 1	Grade 2	Grade 3	Grade 4
Commercial Ethyl Alcohol* %vol./vol.	-	33.0 ±0.5	50.0 ±0.5	50.0 ±0.5
Bensotriasole** %wt./vol.	0.045 – 0.052	0.045 – 0.052	0.045 – 0.052	0.045 – 0.052
Technical Sodium Borate $Na_2B_4O_7.10H_2O$ %wt./wt.	0.5 ±0.1	0.5 ±0.1	0.5 ±0.1	0.5 ±0.1
Colour (a) Dye (b) Concentration %t./vol.	Flourescent Green Sodium Flourescein 0.0005	Green ICI Procrion Brilliant Green H4GS 0.0005	Blue ICI Chlorozal Blue 0.0005	Flourescent Green Sodium Flourescein 0.0005
Water (distilled, demineralised or boiler condensate)	Remainder	Remainder	Remainder	Remainder (after lubricant additive**)

Table 7.3.1 *Hydrolastic fluid detail.*

* This material must be denatured to conform to statutory requirements
** The lubricant for Grade 4 shall be an approved lubricant to give a final mix viscosity of 15.6 to 16.5 sS at 100F. Approved lubricant concentration: Union Carbide Ucon 50HB660 40% vol./vol. Or Union Carbide Ucon 50HB5100 22% vol./vol.

7.3.2 Charging the Hydrolastic Suspension

There are a range of hydrolastic pump units available – from original Churchill style, to home-made items featuring brake/clutch master cylinders, that may be used for charging the hydrolastic system with fluid.

At last resort a pressure-only pump will suffice but it is very important to have some means of evacuating the system. It is not sufficient to just let the suspension down so that the car is resting on the bump stops in the hope that this will force all air from the system. When a hydrolastic displacer is compressed, it does not act like a piston in a cylinder. There is plenty of volume left in the unit and considering that is some cases the units are effectively horizontal, there will be very likely an air pocket in the displacer if the system has been disconnected and fluid drained out of the displacer during repairs.

[21] When BMC Australia engineer Reg Fulford visited Moulton in 1962, he was advised that the alcohol content for fluid used in Australia could be reduced over that intended for UK on the basis of operating temperature range – the clear implication being that the alcohol content acted as an anti-freeze component.

It is very difficult to use a separately plumbed vacuum and pressure pumps. They must be interconnected as per the original Churchill device since it will not be possible to swap from vacuum to pressure functions with separate systems without losing the vacuum.

When evacuating, the vacuum line is attached and then the yellow valve closed and then pumping proceeds until a steady vacuum is obtained. The vacuum has to hold and not leak away after pumping ceases. If the needle on the vacuum gauge starts dropping down to zero, then this indicates a leak that must be found and repaired. Most leaks are to be found in the hydrolastic suspension service unit – usually the fitting that goes on to the Schrader valve on the car. If the hydrolastic service unit has seen a lot of service, then there may be leaks inside the vacuum pump. These can be difficult to repair, but not impossible. Inside this vacuum pump is a piston and cylinder. The piston is a brass assembly with a large, reinforced rubber O ring seal. If the seal is worn, then it can be made to work again by facing off some of the brass assembly in a lathe so that when the clamp nut is tightened, it spreads the seal out a small amount. As well, there are two valves in the bottom of the vacuum unit in the form of Bakelite discs which are easily overlooked and lost. Once a steady vacuum has been achieved, then yellow valve is opened. The vacuum gauge drops to zero. This is normal. What is happening, and what the workshop manual doesn't say, is that in this step, fluid is being sucked into the system from the reservoir. Once the vacuum gauge reads zero, it is best to leave things to settle for about 5 mins so that as much fluid as possible is drawn into the system. At the end of the evacuation procedure, the system is filled with fluid at atmospheric pressure. The vacuum hose can then be disconnected, and the pressure hose attached ready for pressurising the system.

The vacuum pump has two functions. First, air is evacuated from the system. Of course a perfect vacuum cannot be attained, and there will always be some air molecules left in the system, but the objective is to reduce their number as much as possible. Second, after evacuation, the plumbing should be arranged so that hydrolastic fluid can be sucked into the system under atmospheric pressure. That is, the vacuum line cannot be just disconnected, and the pressure pump line attached. Vacuum will be lost as soon as the vacuum line is removed. Using a hand-operated valve, fluid from the storage tank will be sucked into the evacuated system whereupon the vacuum gauge will reduce back to zero. At this point, the system will be filled with fluid at atmospheric pressure. Then the vacuum line is disconnected from the Schrader valve. It is then an easy matter to fit and prime the pressure side of the pump fittings using the T handled valve near the pressure line chuck.

During pumping up, it is important that slow and steady strokes are used with a short pause between each one. This allows the pump cylinder to fill with fluid. If pumping proceeds too quickly, there is a risk that air will be sucked in past the rod seal into the pressure pump rather than sucking in fluid from the tank. The fluid has to find its way past the ball valve into the pump chamber and avoid air molecules being sucked in through the rod seal.

There are of course two connectors that fit on to the Schrader valve. The vacuum connector, or chuck, is a traditional Renrut device made by the Turner company in UK (note the anagram). It only has to hold atmospheric pressure, so it is a simple clamp.

The fluid connector has to hold upwards of 400 psi and so a high pressure connector is required. This has a locking ring that forces to tangs to lock on to the thread to prevent it from flying off under pressure. The actual seal is the upper and lower surface of a cylindrical O ring inside the device. This means that the connector has to be pushed onto the Schrader valve. The downwards pushing of the locking ring also pushes the seal onto the top of the valve and the tangs then hold everything in place. An industrial air chuck with a screw-on fitment is preferable to the standard clamp device.

A bleeder valve is fitted just next to the connector to allow air to be purged from the pressure line before pumping up the suspension itself. Note that the bleeder valve has a small ball bearing at the bottom of it that makes the seal. If the valve is unscrewed all the way, the ball will most assuredly roll away unnoticed.

7.3.3 Ride Height

BMC Australia Engineer Fulford favoured the use of ride height as a means of adjusting the correct amount of pressure in the system. Moulton favoured a pressure reading. Fulford's argument was that any pressure gauge in a typical mechanic's workshop was likely to be broken or inaccurate. And so, we have both methods documented by the factory. Usually the fluid is pumped in until the ride height meets the specification in the workshop manual. However, it is important not to over-pressurise since this will cause the displacer to burst. Having air in the system will result in a very spongy suspension.

The operation of a displacer unit is quite complicated. Think of inflating a flat tyre on your car. Once the rim is supported by the tyre, the whole weight of that corner of the car is taken essentially by the tyre pressure spread over the contact patch. This weight is (for the car just standing there doing nothing) essentially constant. For the tyre, the contact patch is initially quite large. Increasing the tyre pressure causes the contact patch to become smaller. The product of the contact patch area and the pressure remains the same – that is, equal to the "constant" weight of the car at that corner. And so, with hydrolastic systems, because of the conical nature of the "contact patch" offered by the tapered cylinder inside, increasing the pressure decreases the perpendicular area of contact over which the fluid acts so that the force (which equals pressure x area) remains constant. As with a flat tyre, increasing the pressure does raise the car, but the pressure is increasing only because the contact area is getting smaller as the car is raised, the weight remaining constant.

For a Morris Mini with hydrolastic suspension, it is important that the vehicle be loaded with rear seats, rear windscreen, fuel in the tank, battery and spare wheel. Without these items, the rear ride height will be out of specification and much higher than expected when the specified pressure is reached.

Opinions vary as to what pressure one should set these systems to. There are of course the recommendations in the workshop manual. The working pressure for normal driving is usually about 250 psi and should never be increased beyond 300 psi in an attempt to raise the ride height.

7.3.4 The Viscosity of Hydrolastic Fluid

The BMC hydrolastic system came in for a lot of development before being introduced on ADO16, the Morris 1100. BMC were expecting the final units to be able to be disassembled and serviced. Moulton would have none of it, they being sealed for life. when Australian engineer Fulford visited Moulton, his observation was that:

"Moulton was opposed to us doing independent ride work in Australia other than by having modified displacers sent out for trial. He was worried that mistakes and confusion could result. He would also be opposed to changing anything in the valving other than the bleed due to the amount of work that had gone into achieving the present valve design and the complications of changing it."

In fact, Fulford could find very little evidence that proper and systematic testing had been done on the design, it coming to its final state by trial and error.

Despite these worrisome findings, it is evident that the viscosity of the hydrolastic fluid plays some significant part in the overall design and therefore of interest to find out more about this important material property. As benchmarks, the measurements reported here are accompanied by those of water and also ethanol.

According to BMC service literature, hydrolastic fluid contains a lubricity additive to overcome a squeak from the valve inside the hydrolastic unit and the literature lists several different part numbers corresponding to different concentrations of the lubricity additive – the more additive giving a firmer ride. Whether or not the squeaks are reduced due to the lubrication properties of the fluid, or the reduced travel resulting from the change in viscosity is not revealed. Fig. 7.3.1 shows an excerpt from the Austin 1800 Service Training Notes TP765.

The charging of this system is carried out through a Schrader valve in the pipe line. The system is charged with hydrolastic fluid, which is an anti-freeze solution of constant viscosity, containing a mixture of demineralized water and alcohol, a rust inhibitor and an agent that is added to make the fluid distasteful - a legal requirement.

Fig. 7.3.1 *The nature of hydrolastic fluid.*

Fulford discussed this with the UK engineers and reports: *"we should be able to reduce the alcohol content of the fluid for our climate. I said I would discuss this with Brown."* – the implication being that the alcohol content was purely for antifreeze purposes.

For the purposes of this Section, three fluids were tested in an instrument designed to measure absolute values of viscosity and elasticity, or shear modulus. To understand the significance of the readings, it is necessary to know a little but about these physical properties.

In solid mechanics, it is the elastic modulus of material that is of most interest, as well as other properties such as yield strength, hardness, and fracture toughness. In fluid mechanics, it is the viscosity, or "resistance to flow" that is of most interest. However, many materials have properties somewhere in between being solid and liquid and are often referred to as being "visco-elastic".

A good example of a purely viscous material is water. When water is poured, it flows and takes the shape of its container. Compare this with something like cottage cheese. This material also flows (if say deformed with a butter knife), but it has "body" to it. It will sit there, an elastic quivering mass, unless it is deformed.

It is the shear modulus *G* that quantifies the elastic component of the material response and the viscosity η that quantifies the viscous response. Water theoretically has $G = 0$ and a viscosity $\eta = 1$ mPas at room temperature. These properties often change with temperature, and also the rate of deformation. For example, manufacturers of paint arrange things so that paint is easy to brush (low viscosity when the shear rate is high) but has a high viscosity when it is left alone (so that it doesn't sag and cause runs).

Fig. 7.3.2 shows the viscosity η and shear modulus *G* on the vertical axis against frequency of deformation on the horizontal axis. Note that hydrolastic fluid has a higher viscosity than water, which in turn has a higher viscosity than ethanol. For the shear modulus, hydrolastic fluid behaves very much like the other fluids which are all close to zero.

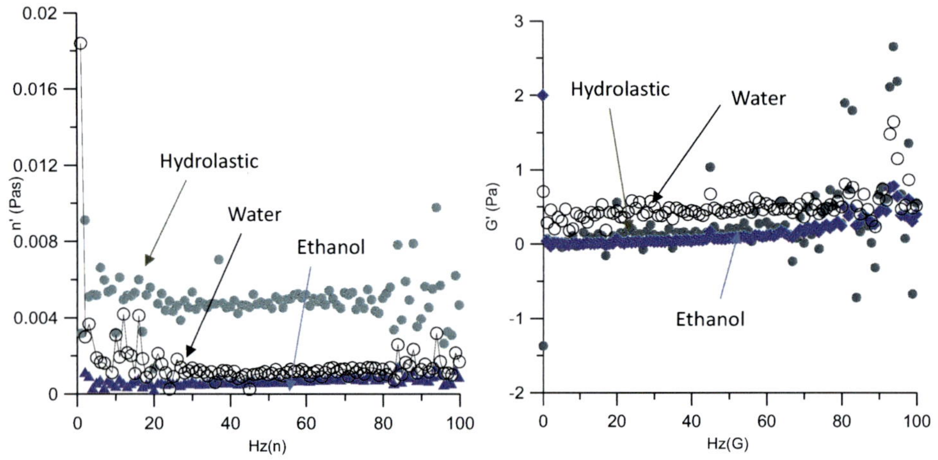

Fig. 7.3.2 *Viscosity measurements*

The results show that the shear rate (in Hz) does not influence the properties very significantly and so the characteristics of a hydrolastic suspension are likely to be the same over bumpy roads and smooth roads.

Hydrolastic fluid is not simply a mixture of water and alcohol as evidenced by the significantly higher value of viscosity throughout the frequency range. It is the lubricity additive that makes the difference here and is why the fluids with the high concentration of additive give a harder ride since the fluid does not flow from front to rear as easily as would water. Hydrolastic fluid does indeed feel quite oily when rubbed between the fingers.

7.3.5 Austin 1800 Torsion Bars

One of the more unusual features of the Australian-designed Austin 1800 utility is the torsion bar arrangement at the rear suspension.

Because the vehicle is designed to carry a payload, the rear displacers were increased in size to match those at the front – unlike the saloon which used smaller diameter displacers at the rear.

In addition to larger diameter hydrolastic units, a torsion bar was also fitted to each of the rear suspension arms. Although we might at first think that the torsion bars serve to assist the hydrolastic springs in carrying the payload, an analysis of the way in which they are set reveals that much more is going on.

In this article, the action of the torsion bars and the way in which they are set is described in detail so as to reveal their true function.

7.3.5.1 Setting Procedure

The figure below shows the general arrangement and factory instructions for torsion bar initial setting.

With the hydrolastic pressure released, the rear radius arm is positioned so as to give the clearance shown 5" from the centre of the wheel bearing to the load floor. This is the torsion bar neutral position. The torsion bar brackets are then adjusted for no torsion (unstressed) and locked into position.

The hydrolastic system is then pressurized in the normal manner.

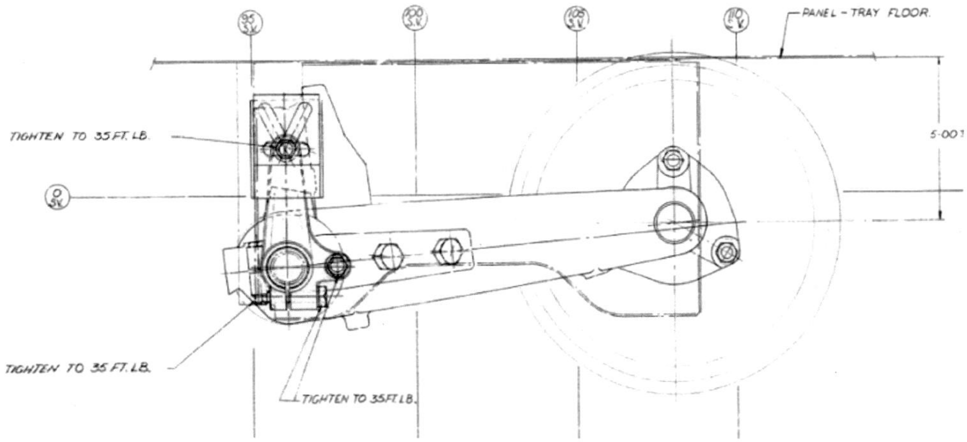

TORSION BAR SETTING INSTRUCTIONS.
INITIAL PRODUCTION AND RECTIFICATION.

1. WITH THE HYDROLASTIC SUSPENSION SYSTEM DEPRESSURISED, SUPPORT THE RADIUS ARMS AT THE HEIGHT SHOWN ABOVE, WHICH IS THE NEUTRAL POSITION OF THE TORSION BAR. (IE. THE TORSION BAR IS UNSTRESSED.)
2. FIT VERNIER BRACKET TO TORSION BAR SPLINES IN A SUITABLE ANGULAR POSITION, WHICH WILL ALLOW FITMENT OF THE CLAMPING BOLTS. (VERNIER BRACKET TO BODY).
3. FIT LOCKING PLATES, ALL BOLTS ETC. AND TIGHTEN. NOTE THAT LOCKING PLATE CAN BE TURNED OVER AND/OR MOVED UP AND DOWN TO ACCOMMODATE ANY POSITION OF THE TOP VERNIER BRACKET BOLT.

SERVICE.

1. WITH THE TORSION BAR VERNIER BRACKET REMOVED, LOAD THE VEHICLE AT THE REAR OR ADJUST HYDROLASTIC PRESSURES (WHICHEVER IS APPLICABLE) TO POSITION THE RADIUS ARMS AS SHOWN. THIS IS THE NEUTRAL POSITION FOR THE TORSION BAR.
2. AS FOR INITIAL PRODUCTION.
3. AS FOR INITIAL PRODUCTION.

Fig. 7.3.3 *Setting procedure for torsion bar.*

But, when the hydrolastic suspension is pressurised, the clearance above is considerably greater than 5" – meaning that at a zero load condition, the hydrolastic rubber spring is being compressed by the torsion bar. This is a pre-stress that is applied to the rubber spring before any vehicle load or payload is applied.

One would have thought that the purpose of the torsion bar was to assist in taking the load, but instead, as far as the rubber spring is concerned, the torsion bar prestress adds to the load it had to bear - well, up to a certain point at least. Eventually, with increasing load, the torsion bar will be twisted back to its neutral position, and from then on, it will assist the rubber spring in taking the load. One might ask the question why is this so?

7.3.5.2 Analysis – Linear Spring

To understand the action of the torsion bar, we might model two cases. The first being the action of a torsion bar and a rubber spring both with a linear characteristic. In fact, the rubber spring has a non-linear response and we will get to that case shortly, but let's start with a simple case of two linear springs, interconnected, one acting in compression and the other pre-tensioned as shown in the figure below.

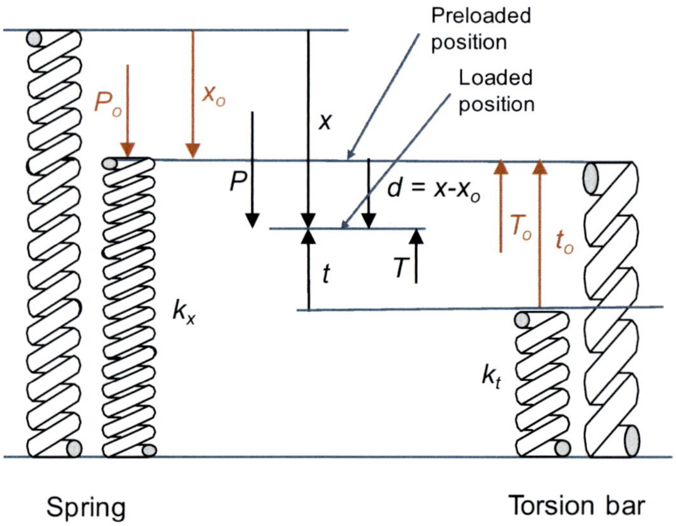

Fig. 7.3.4 *Two springs in parallel, one the "rubber" compression spring, the other, the torsion bar.*

For convenience, we will consider the weight of the vehicle as the "payload" P such that at zero load, the only force acting on the rubber spring (hereafter called the "spring") is the pre-tension T of the torsion bar. We will then examine what happens when load is applied (and this load could be the weight of the vehicle, plus any additional load that might be carried in the tray).

In the analysis to follow, "up" is taken as the positive direction, and "down" as the negative direction. The deflection of the spring is taken from the position at its free length.

As per the instructions shown in Part 1, the operator effectively sets an initial preload T_o.

Since in this case, both springs are linear springs, then the following equations apply – where via Hooke's Law, the load is linearly proportional to the displacement x of the end of the spring via spring constants k_s and for the displacement t of the torsion bar, k_t.

$$P_o = k_s x_o$$

$$T_o = k_t t_o$$

Since at this condition, the only load applied to the spring P_o is supplied by the "tension" T_o of the torsion bar such that:

$$P_o = -T_o$$

Thus,

$$x_o = -\frac{k_t}{k_s} t_o$$

At some payload P, the total force F now acting on the spring is the sum of the P and that from the torsion bar T.

$$F = P - T$$

$$= k_s x$$

And

$$T = k_t t$$

where

$$t = t_o + (x - x_o)$$

And so,

$$k_s x = P - k_t\big(t_o + (x - x_o)\big)$$

Collecting terms and making x the subject of the formula, we have the deflection of the spring as:

$$x = \frac{P - k_t t_o + k_t x_o}{k_s + k_t}$$

From the initial condition (pre-stress), the load point displacement for the payload P is thus:

$$d = x - x_o$$

Fig. 7.3.5 below shows the characteristics of the spring response for different scenarios. The numbers shown are arbitrary, but the vertical axis represents force, and the horizontal axis, displacement of the free end of the spring.

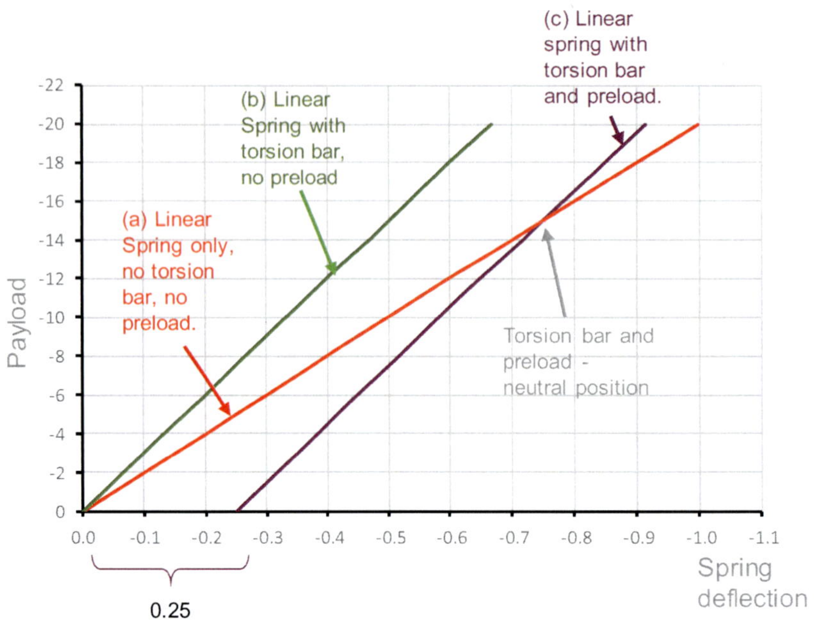

Fig. 7.3.5 *Spring response for (a) Linear Spring, no Torsion Bar and no preload, (b) Linear Spring with Torsion Bar – no preload, (c) Linear Spring with Torsion Bar and preload.*

As would be expected, (a) and (b) show the increase in stiffness of the combination when the torsion bar is included. The existence of a preload serves to shift the response to the right, but the slope of the response is the same as before.

Note that as P increases, at some corresponding displacement, the torsion bar reaches its neutral position and at that point, the load applied to the spring is the same at which would occur for the spring alone, the torsion bar contributing nothing to the load applied to the spring.

The conclusion here is that although the torsion bar stiffens the suspension, the preload does not make any difference and only serves to lower the height of the vehicle a constant amount.

7.3.5.3 Analysis – Non-Linear Spring

We now come to the more interesting case of a non-linear spring. This is more realistic for the hydrolastic suspension system used in the vehicle since the rubber, and the shape of the cone at the end of the pushrod, effectively stiffens the response as a function of displacement. That is, the spring "constant" is a function of displacement. We might represent this as a square law such that:

$$P = k_s x^2$$

The torsion bar remains as a linear spring.

At the initial preload state, then:

$$P_o = k_s x_o^2$$

$$T_o = k_t t_o$$

As before, the only load applied to the spring P_o is supplied by the "tension" T_o of the torsion bar such that:

$$P_o = -T_o$$

Thus,

$$x_o = -\sqrt{\frac{k_t}{k_s}}t_o$$

At some payload *P*, the total force *F* now acting on the spring is the sum of the *P* and that from the torsion bar *T*.

$$F = P - T$$

$$= -k_s x^2$$

The minus sign is necessary here because the downwards direction (*x*) is negative as is the force *F*.

Also,

$$T = k_t t$$

Whereas before:

$$t = t_o + (x - x_o)$$

And so,

$$k_s x^2 = P - k_t\big(t_o + (x - x_o)\big)$$

Collecting terms and making *x* the subject of the formula, we can solve the resulting quadratic equation to obtain:

$$x = \frac{-k_t - \sqrt{k_t^2 - 4(-k_s)x_o(-k_t t_o + k_t x_o)}}{-2k_s}$$

In this equation, a minus sign in put before k_s to account for the negative direction of both *x* and *F* as said above.

From the initial condition (pre-stress), the load point displacement for the payload P is as before:

$$d = x - x_o$$

Fig. 7.3.6 below shows the characteristics of the spring response for different scenarios including a comparison with the case of a linear spring (Case (c) as before).

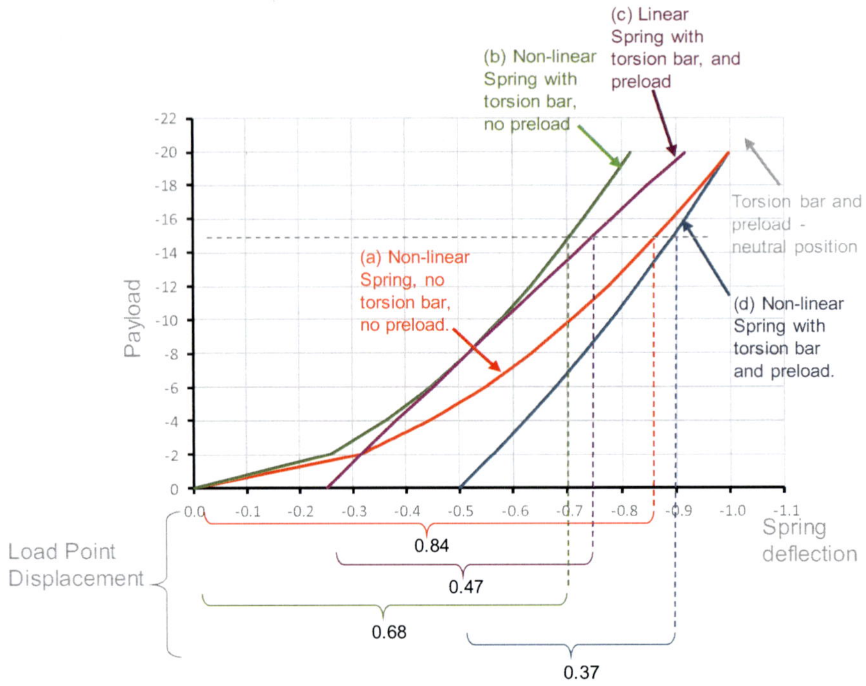

Fig. 7.3.6 *Spring response for (a) Non-Linear Spring, no Torsion Bar and no preload, (b) Non-Linear Spring with Torsion Bar – no preload, (c) Linear Spring with Torsion Bar and preload (d) Non-Linear spring with torsion bar and preload.*

Also shown above is the load point displacement for each scenario for a selected arbitrary load. As can be seen from Fig. 7.3.6, for a given arbitrary load of 14 force units, the load point displacement is the least for the case of a non-linear spring with a preloaded torsion bar. This would indicate that the overall stiffness of the spring-torsion bar system is stiffer compared to the other cases.

7.3.5.4 Discussion and Conclusion

In the case of a linear spring and torsion bar, the presence of an initial preloading made no difference to the stiffness of the spring combination. In the case of a non-linear spring and torsion bar, the presence of a preload increases the stiffness of the spring system as evidenced by the lower value of load point displacement for a given load. This occurs because the initial preload point of the spring system is set further up into the stiffer part of the rubber spring non-linear response.

The actual shape of the rubber spring characteristic is not modelled since this information is not available, and so a simple square law is treated for the sake of illustrating the effect.

The purpose of setting a preload is to take advantage of the non-linearity of the rubber spring in the hydrolastic unit that reducing the change in attitude of the vehicle between the loaded and unloaded conditions. This is necessary in the case of an Austin 1800 Utility with interconnected hydrolastic suspension since the headlight aim would otherwise be unacceptable – a circumstance already in evidence in the saloon (with smaller rear displacers) when carrying rear seat passengers and luggage.

7.4 Secret Life of a Tyre

The mechanism by which a tyre supports the weight of a motor vehicle is a somewhat complicated one. When the tyre air pressure is low, the tyre appears flat, and the weight of the car makes the whole affair sag so the air pressure does have something to do with it. But how is the weight of the car supported by this air pressure if the pressure is the same all the way around the rim? Surely the air pressure pushing down on the top half of the rim surface will be balanced out by the pressure pushing upwards from the bottom half? So exactly how is the weight supported?

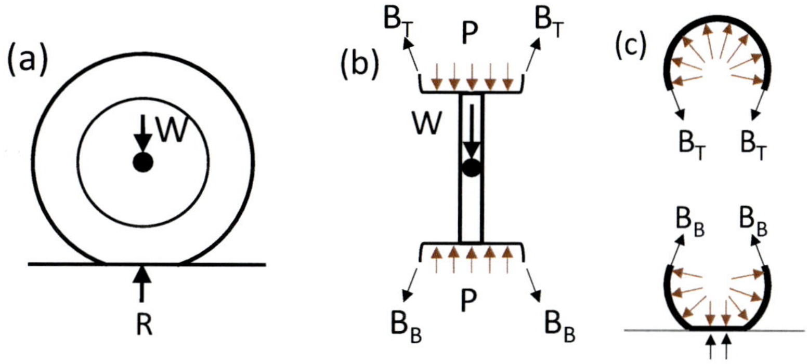

Fig. 7.4.1 *(a) Weight W of car is balanced by reaction R from road. (b) Free body diagram of wheel (c) Free body diagram of tyre.*

Consider Fig 7.4.1 (a). Here is shown the weight of the car *W* and the reaction force *R* from the road. The two forces are in balance (otherwise the wheel and tyre would move in the vertical direction). The question is, how is the force *W* at the wheel hub transmitted down to the surface of the road so that the road can support it?

Pressure in a gas is distributed evenly throughout the volume. That is, in a balloon full of compressed air, it is not found that the pressure in the balloon is greater at one place compared to another. Rather, the pressure is uniform throughout. Now, when pressure acts over a surface, the force applied is found by multiplying the pressure times the area over which the pressure acts.

What this means for a tyre and wheel is that the pressure inside the tyre is distributed uniformly throughout the volume of the tyre. So, for every square centimetre of wheel rim surface, we have the same force acting. Further, this force acts perpendicular to the surface upon which it acts. But, if the pressure is the same at the top and the bottom of, and all the way around, the rim, then all these forces will balance out. So, the question is, if the forces due to the air pressure balance out, just what is supporting the weight *W* of the car?

To appreciate what is happening, a free-body diagram of the cross-section of the wheel is required. This is shown in Fig. 7.4.1 (b). The wheel is drawn in complete isolation except that each and every force acting on the object is shown. In this case, the forces acting directly on the wheel are:

(1) The weight *W* of the car.
(2) The distributed forces *P* due to the pressure of the compressed air.
(3) The forces *B* acting on the rim where the tyre bead is in contact with the rim.

Since the forces due to the air pressure completely balance out, these need not be considered further. What is left are the forces acting at the bead. What are these forces? Well, it is easy to imagine that when the tyre has pressure in it, the fabric of the tyre is placed under tension, just like the surface of a balloon is under tension when it is inflated. That is, the air pressure pushes outwards on the tyre walls and the side walls are placed in tension in the radial direction (forgetting about tensile forces around the circumference at the moment).

The free body diagram of the *tyre* in cross section is now to be considered as shown in Fig. 7.4.1 (c).

The tension in the side walls depends on the air pressure inside the tyre. But, and here is the crucial point, the downwards pressure from within the tyre over the contact patch is balanced by the upwards pressure over the contact patch from reaction force from the road. That is, the pressure over the contact patch is not contributing to the tension in the radial plies in the side walls because the tyre material is sandwiched between the inside pressure and the road surface.

The effect can be readily demonstrated by putting the palm of your hand flat against the side of a party balloon. The material under your palm relaxes as the tension in that part of the balloon surface is relieved.

This means that for a tyre, the tension in the side walls at the bottom is less than the tension force in the side walls at the top. What this then means is that there is a greater tensile force in the ply at the top of the tyre than there is at the bottom. It is this difference in tensile force that supports the weight of the car.

It also means that even when a wheel and tyre is rolling along in a perfectly straight line, the tread is wearing as a result of the friction generated as a result of the shrinking of the contact patch inwards both longitudinal and transverse directions.

Consider a couple of different cases:

1. The tyre pressure is reduced: In this case, the tyre would go slightly flat and the contact patch area would increase. But the *difference* between the tension in the top and bottom of the radial plies would remain the same, even though the absolute value of these tensile forces would be less. It is the difference in tension that is important.
2. The tyre pressure is increased: In this case, the contact patch area would decrease, and less of the radial ply would be sandwiched between the road and the air pressure force. In other words, more of the radial ply of the tyre at the bottom would be under tension, but the *difference* in tension between the top and the bottom would remain the same.
3. The weight of the car is increased: the contact patch increases in size and so the *difference* in tension between the top and the bottom plies is increased by the same amount.

So, contrary to expectations, the wheel is actually hanging inside the tyre bead from the extra tensile force in the radial plies at the top.[22]

[22] The author thanks Mr Brian Le Grice, of Sydney Technical College, for the remarkable explanation of this commonplace event.

CHAPTER 8.
STEERING

8.1 The Ackerman Principle

8.1.1 Rolling Motion

It is instructive to examine the design of the steering mechanism of a motor vehicle so that an appreciation of the underlying issues can be gained. Rather than discuss the well-known parameters of castor, camber and so on, in this section, we look at some more fundamental issues.

When a vehicle is moving in a straight line, the wheels have a purely rolling motion in the direction of motion. When it is desired to have the vehicle turn, then in order to preserve a purely rolling motion and thus avoid any slip at the contact patch between the tyre and the road, it is necessary that the wheels take a circular path whose centres all meet at one point. In a simple beam axle type of steering, such as may be found on a billy cart, the situation is as shown in Fig. 8.1.1:

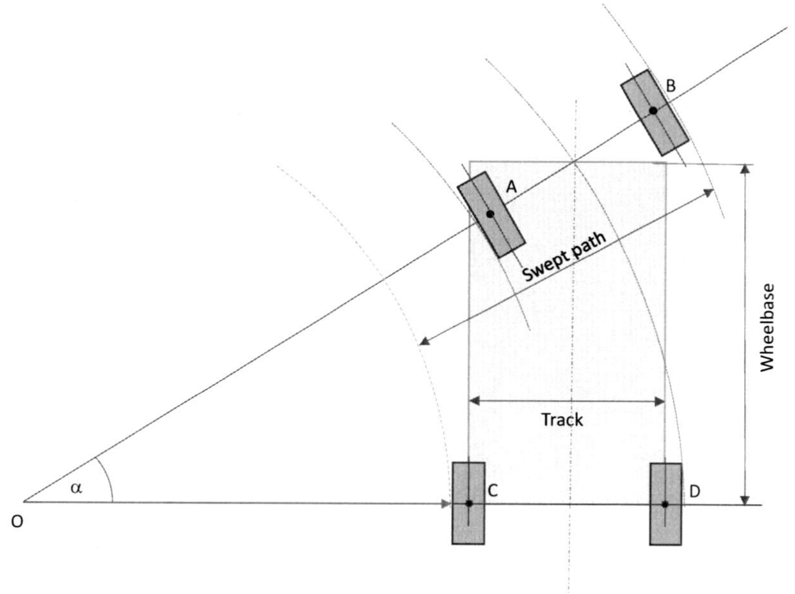

Fig. 8.1.1 *Billy cart steering mechanism. All wheels turn about a fixed point O.*

This type of steering is unsuitable for motor vehicles on account of the excessive lateral movements of the wheel and bub, and so an alternative approach is usually taken.

8.1.2 Perfect Turning

Rather than pivot the whole front axle, each wheel pivots on a stub axle such as shown in Fig. 8.1.2. For the wheels to all turn about a single point O, the outer front wheel must be at a different angle to the inner wheel.

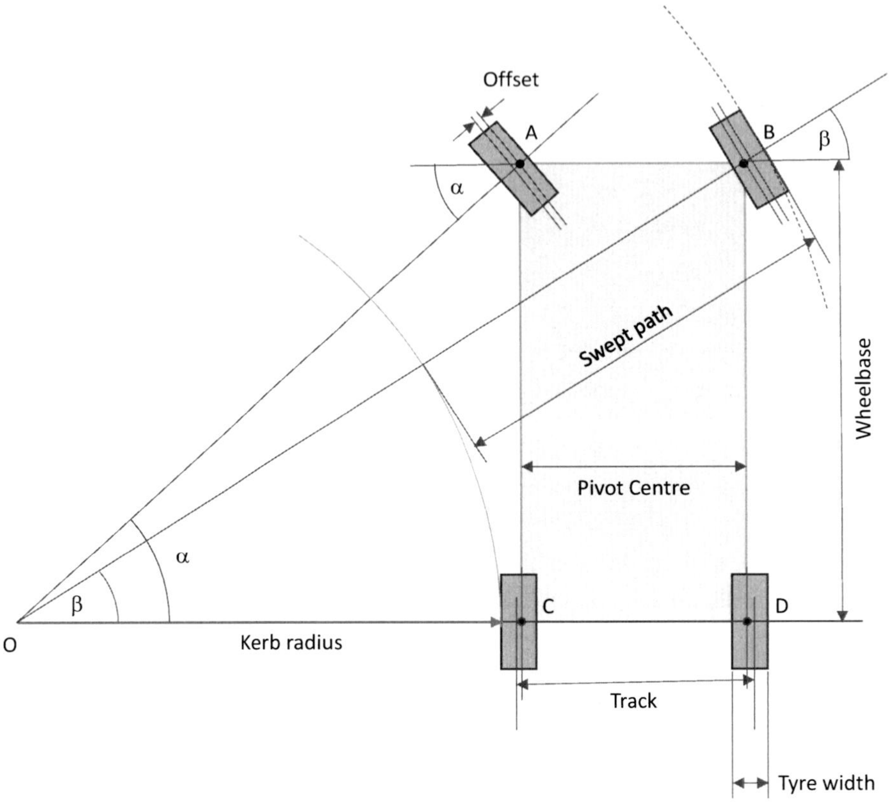

Fig. 8.1.2 *Stub axle steering. All wheels turn about a fixed point O. Dimensions shown for Morris Mini.*

For a given vehicle dimension, AC is the wheelbase. OC is the kerb radius plus half the tyre width and the offset. We may wish to know, for example, what the angle β should be for a particular turning α of the inner wheel given the track and wheelbase.

With reference to Fig. 8.1.2, the following geometrical analysis applies:

$$tan(\alpha) = \frac{AC}{OC}$$
$$OC = \frac{AC}{tan(\alpha)}$$

Now, with respect to the angle β,

$$tan(\beta) = \frac{BD}{OD}$$
$$= \frac{BD}{OC + CD}$$

But CD is the pivot centre P, and AC and BD is the wheelbase W. Thus,

$$tan(\beta) = \frac{W}{\frac{W}{tan(\alpha)} + P}$$

$$\frac{1}{tan(\beta)} = \frac{1}{tan(\alpha)} + \frac{P}{W}$$

The pivot centre P is:

$$P = Track - 2(Offset)$$

Note that as long as the relationship between α and β is maintained as shown above, then the wheels will all turn about the centre point O, for all kerb radii.

For a particular value of steering angles α and β, the kerb radius is:

$$KR = \frac{W}{tan(\alpha)} - Offset - \frac{Tyre\ Width}{2}$$

and the swept path is found from:

$$sin(\beta) = \frac{BD}{OB}$$

$$OB = \frac{W}{sin(\beta)}$$

where

$$SP = \left[\frac{W}{sin(\beta)} + Offset + \frac{Tyre\ Width}{2}\right] - Kerb\ Radius$$

The quantity in square brackets is the radius of the turning circle.

As can be seen from Fig. 8.1.2, the swept path is considerably larger than the track and increases as the steering angles increase, thus the requirement that long wheelbase vehicles be given plenty of room to turn.

For example, say our vehicle dimensions are shown as in Table 8.1.1 and the inner wheel is turned $\alpha = 15°$. Calculations show that the resulting value of $\beta = 13°$ and this would result in a kerb radius of 7.5m and a swept path of about 1.5m.

8.1.3 Four-Link Ackerman Geometry

A practical method of achieving close to perfect turning geometry is achieved using angled steering arms, whose line of intersection is shown traditionally at or near the midpoint of the rear axle as shown in Fig. 8.1.3.

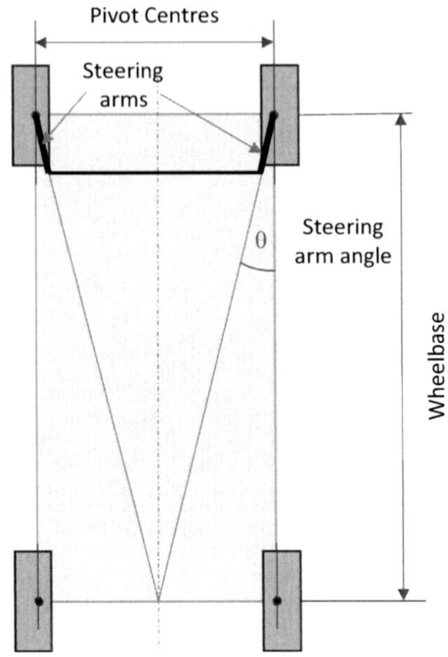

Fig. 8.1.3 *Traditional arrangement of angled steering arms.*

This geometry is known as Ackerman[23] steering. It provides true rolling action in the straight ahead position, and in most circumstances, one angle to the left or right. At other steering positions, the axes of the front wheel do not move around the centre of movement of the rear axle, and there is a certain amount of slip. The angle at which perfect rolling action occurs is determined by the geometry of the track, wheelbase and steering arm angle.

Such geometry is amenable to geometrical analysis, and it is worthwhile us doing so. Let us first analyse a conventional four link system where it is imagined that the points *a* and *b* in Fig. 8.1.3 are effectively linked together by a beam with stub axles free to rotate at each end via interlinked steering arms of length *l* and tie rod M.

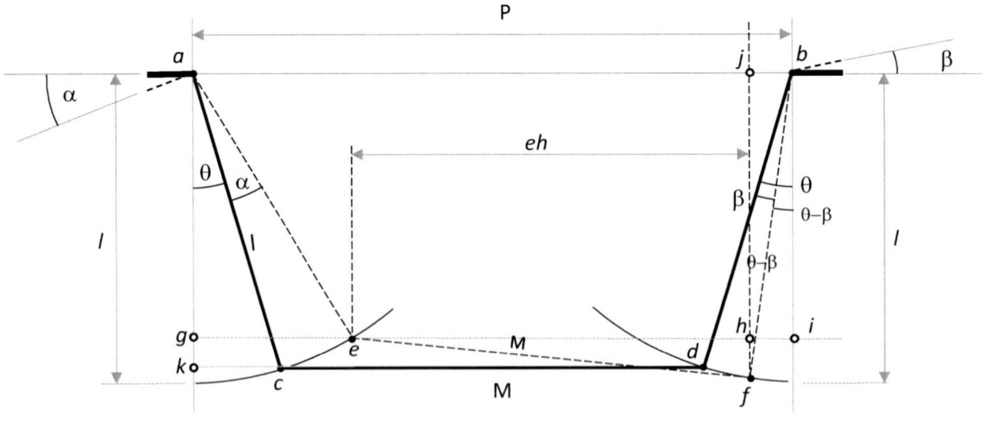

Fig. 8.1.4 *Geometrical analysis of four-link steering mechanism.*

[23] Patented by Rudolph Ackermann (to use the German spelling) in 1818.

Geometrical considerations show that on the left hand side:

$$ga = lcos(\alpha + \theta)$$
$$ge = lsin(\alpha + \theta)$$
$$kc = lsin(\theta)$$

And on the right hand side,

$$jh = ga$$
$$jf = lcos(\theta - \beta)$$
$$hi = jb = lsin(\theta - \beta)$$
$$hf = jf - jh$$

and

$$eh = \sqrt{M^2 - hf^2}$$

where

$$T = M + 2kc$$
$$\therefore M = T - 2lsin(\theta)$$

Thus, the Pivot Centre distance *P* is the addition of the lengths *ge*, *eh* and *hi*:

$$P = ge + eh + hi$$
$$= lsin(\alpha + \theta) + \sqrt{\left(P - 2lsin(\theta)\right)^2 - \left(lcos(\theta - \beta) - lcos(\alpha + \theta)\right)^2} + lsin(\theta - \beta)$$

Despite the awkward expression, this equation allows us to compute the angle β for a particular angle α given the dimensions of the four-link steering geometry, namely *P*, *l* and θ. Since β is not the subject of the formula, a trial value may be set with a given value of *P* and the resulting computed value of *P* compared, and β adjusted accordingly. If α and β happen to be related by the equation given in Section 8.1.2, then we can compute the kerb radius for perfect turning should we wish.

8.1.4 Rack and Pinion

For rack and pinion steering, the situation is more complicated than the four link arrangement since the tie rods add additional pivoting members to the motion. For the present analysis, it is assumed that the rack is neither set back or forward from the steering tie rod ends – that is, the tie rods are in line with the rack at the straight ahead position.

Geometrical considerations show that for Fig. 8.1.5, on the left hand side:

$$ga = lcos(\alpha + \theta)$$
$$ge = lsin(\alpha + \theta)$$
$$kc = lsin(\theta)$$
$$ka = lcos(\theta)$$
$$gk = ka - ga$$

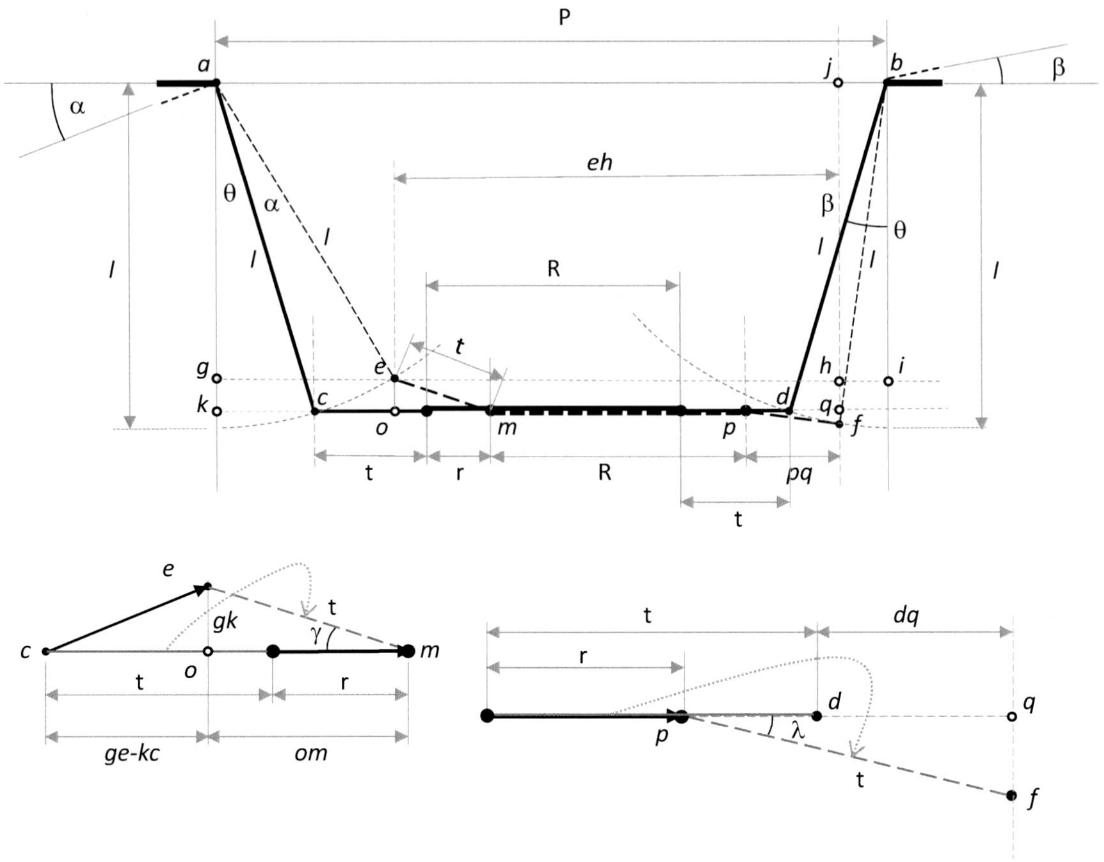

Fig. 8.1.5 *Rack and pinion steering geometry.*

For the triangle Δ_{cem},:

$$sin(\gamma) = \frac{gk}{t}$$
$$\gamma = sin^{-1}\left(\frac{gk}{t}\right)$$
$$om = \frac{gk}{tan(\gamma)}$$

And so, the rack movement *r* is:

$$r = (ge - kc) + om - t$$

On the right hand side:

$$jh = ga$$
$$jf = lcos(\theta - \beta)$$
$$hi = jb = lsin(\theta - \beta)$$
$$jq = ka$$
$$qf = jf - jq$$
$$pd = t - r$$

For the triangle Δ_{pqf}:

$$sin(\lambda) = \frac{qf}{t}$$

$$\lambda = sin^{-1}\left(\frac{qf}{t}\right)$$

$$tan(\lambda) = \frac{qf}{pq}$$

$$pq = \frac{qf}{tan(\lambda)}$$

And so:

$$eh = om + R + pq$$

The Pivot Centre distance *P* thus is the addition of:

$$P = ge + eh + hi$$

While somewhat inconveniently expressed as *P*, which is usually fixed by other design criteria, other parameters may be adjusted until the computed value of *P* meets the required value. In the case of the four-link geometry, the distance *M* in Fig. 8.1.4 is determined by the Pivot Centre distance *P*, whereas for the rack and pinion geometry, the corresponding distance is set by the choice of tie rod length and rack length.

8.1.5 Graphical Methods

As can be see above, even for the most simplest of arrangements, analytical methods can be cumbersome. Often as not, steering angles are described by graphical construction by the drawing office as shown in Fig. 8.1.6 for ADO17 Austin 1800. Recall our brief mention of these methods in Chapter 2 in relation to the mechanics of machines.

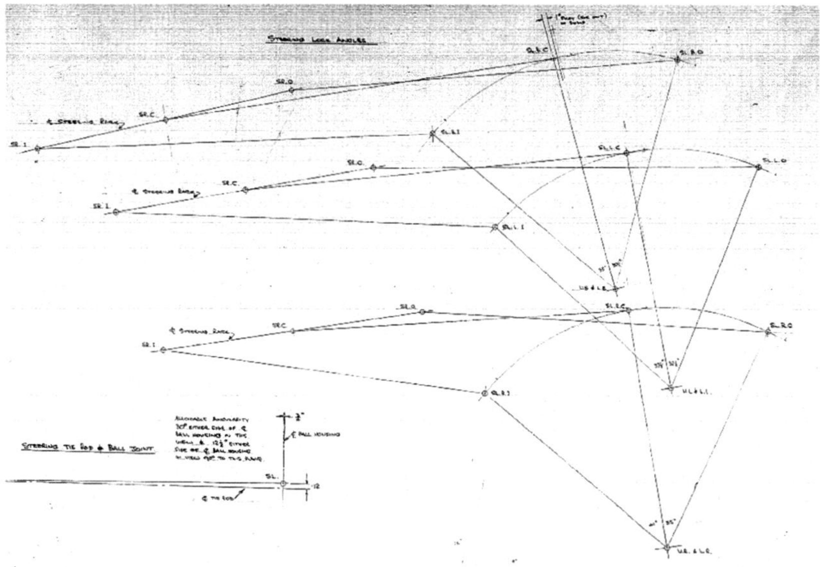

Fig. 8.1.6 *Graphical analysis of Ackerman steering geometry ADO17 (EXP2884).*

At the time of the P86 development in 1972, Leyland Australia had developed a computer program in Fortran on an IBM360 to compute the roll centres, front wheel toe-in and toe-out, camber, caster, coordinates of the lower arm ball joint, tyre contact patch, wheel centre, and steering wheel tie rod ball joint centre with suspension movement for various positions of the rack position.

8.1.6 Perfect Turning: Morris Mini

Due to the large leverage ratios involved, any calculations in relation to suspension geometry prove extremely sensitive to minor variations in input parameters. The geometry for a Morris Mini is particularly sensitive due to the small wheelbase relative to the track. In this section we shall concentrate on the rack and pinion geometry but will compare with the four-link layout for comparison. The factory data for a Morris Mini are shown in Table 8.1.1.

Wheelbase (W)	2032	mm
Track (T)	1205	mm
Camber	2	deg
Castor	3	deg
KPI	9.5	deg
Toe-out (straight ahead)	1.6	mm
Toe-out on turn (full lock)	23 ±1.5°(α), 20°(β)	deg
Tyre Width	98.6	mm
Tie rod length (t)	224 (228.5)	mm
Rack length (R)	600	mm
Lock to lock	2.3	Turns
Rack Travel (left or right)	44.5	mm

Table 8.1.1 *Steering and suspension geometry data Morris Mini.*

We would like to know at what angle α does perfect turning occur, and consequently, the kerb radius and steering wheel position at that angle.

To answer this question, it is necessary to know the steering arm length and the steering arm angle. Although the actual steering arm has a bent appearance, it is the relationship between the ball socket in the end of the tie rod to the steering axis that is important. The actual shape of the arm is immaterial, and indeed, the drawings show that not only is the arm bent, but it is tilted to the horizontal as fitted to the vehicle.

To determine the turning circle, and the pivot centre distance, we require the offset, which can be obtained graphically and is shown in Fig. 8.1.7 as 30mm.

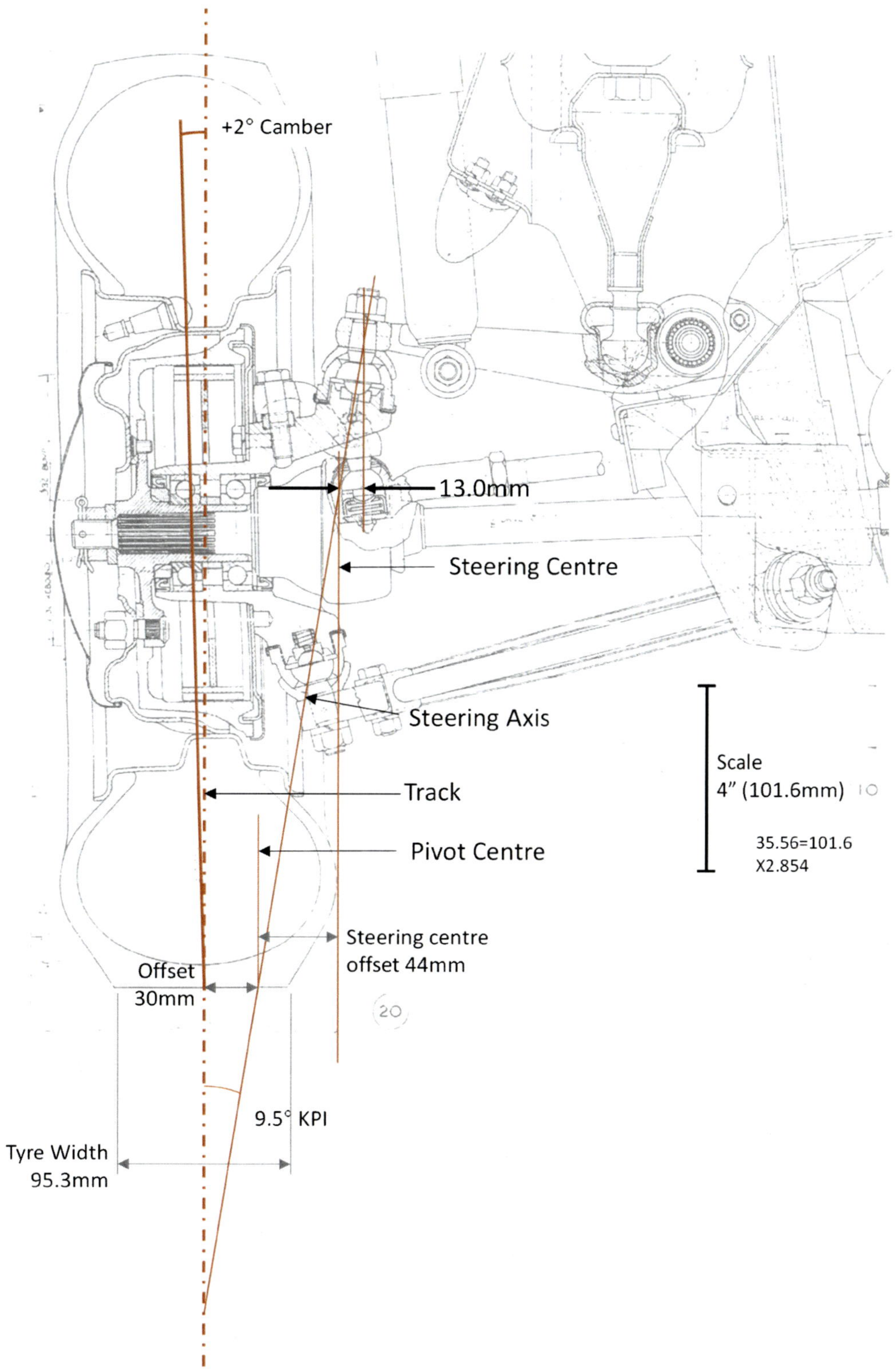

Fig. 8.1.7 *Steering detail, Morris Mini.*

Chapter 8. Steering 165

Fig. 8.1.6 is a very simplified description of the actual geometry. The point on the road at which the wheel is seen to turn is the pivot centre, usually somewhat inboard of the track (the latter taken as the distance between the centres of the tyre widths). Fig. 8.1.6 assumes that the steering axis is vertical and is aligned with the pivot axis. However, due to king pin inclination, the hub as an assembly rotates about an inclined steering axis as shown in Fig. 8.1.6. Graphical analysis shows that the steering centre (the point at which the steering arms actually rotate) is found at the intersection of the steering arm ball joint with the steering axis. This is shown as 13mm on Fig. 8.1.7. Because of this, the pivot centre on the road rotates around an arc as the steering arm turns. While amenable to analytical methods, for our purposes here, we will make a simplification.

From the point of view of the contact with the road, the steering arm angle is taken to be that which is subtended by the distance labelled Steering Centre Offset shown in Fig. 8.1.7 and is shown as 44mm. For the actual steering arm, scaling off the drawing in plan view allows us to estimate a value for the rearward projection as shown in Fig. 8.1.8 and is shown as 98.6mm. These dimensions give a steering arm angle of 24.4° and length 106.4mm.

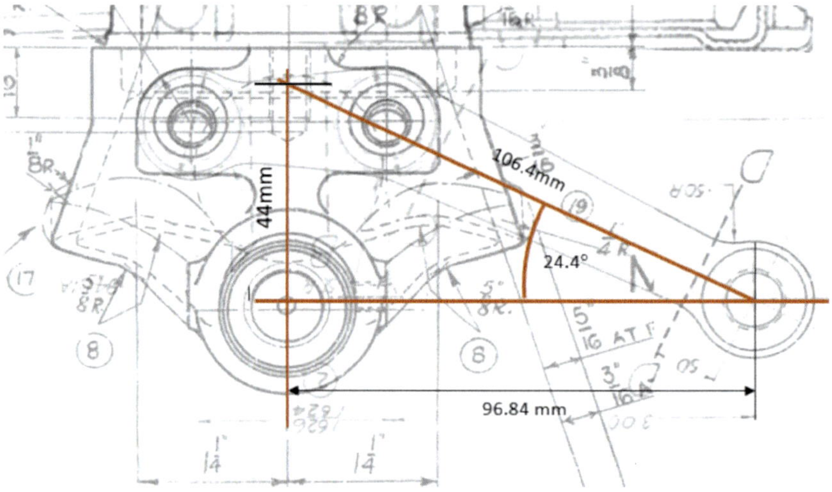

Fig. 8.1.8 *Steering arm and swivel hub, Morris Mini.*

One last dimension that is important is the tie rod length. The steering rack length is specified on the drawing at 600mm, but the tie rod lengths are adjustable. In the case of the four-link geometry, the effective length of M is the distance between the ends of the steering arm ball joints, so for the calculations here, we have set the tie rod length to achieve this dimension with the fixed rack length. This gives us perfect turning at the straight ahead position.

To obtain an appreciation of the situation, it is necessary to examine the difference between the actual computed geometry and that which gives perfect turning over a range of angle α. Fig. 8.1.9 shows the results of computations with $\theta = 24.4°$ and $l = 106.4$mm, and the data shown in Table 8.1.1. The results for the four link analysis are shown for comparison.

For the rack and pinion steering, it is immediately obvious that perfect turning occurs over a very flat range from the straight ahead position for turning angles up to about 12° and thereafter there is an increasing deviation up to full lock at about 25° where we have stopped the calculations. The vertical scale in Fig. 8.1.9 has been magnified for clarity, but the deviation even at α = 25° is only of the order of 1.4%.

This is a very desirable state of affairs since it means that this vehicle has almost perfect steering geometry over the range of steering wheel turn most commonly encountered at moderate to higher speeds, whereas for sharper turns, usually made at parking speeds, we do not care overly much if the tyres scrub a little, and even then, the error in outer wheel angle from perfect turning is less than 2%.

The point of intersection of the steering arms *b* is computed to be 1194mm behind the front wheels at about 59% of the wheelbase. This dimension is found from:

$$b = \frac{P - 2(Offset)}{2tan(\theta)}$$

where *b* is shown graphically in Fig. 8.1.10.

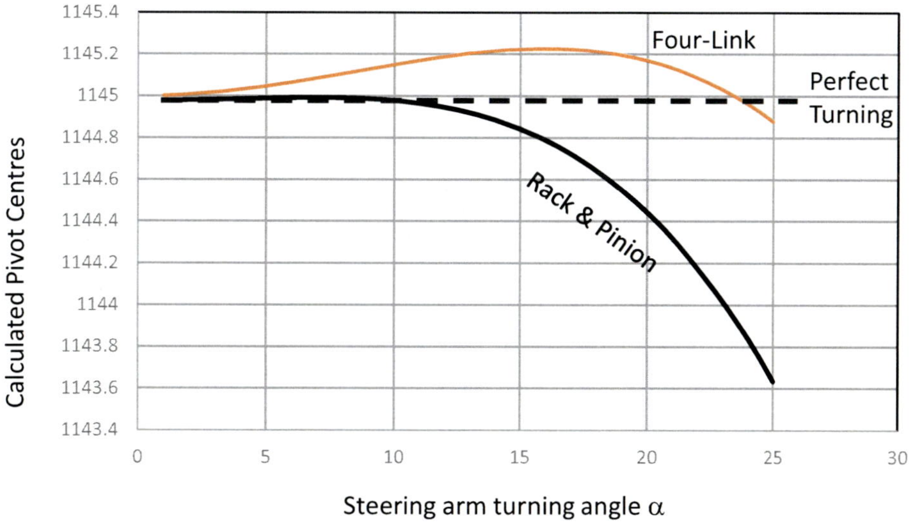

Fig. 8.1.9 *Error curve showing points at which perfect turning occurs ADO15.*

At α = 23°, calculations (assuming perfect turning) show a turning circle of 12.7m, a swept path of 1.65m, and about 0.9 turns of the steering wheel. Comparison with the factory data for α = 23° indicates that in practice, the outer wheel doesn't turn quite as much as is necessary for perfect turning. On the other hand, computations show a lower number of turns, and a larger turning circle compared to the factory data indicating that the 23°/20° toe-out on turns data is not at full lock as is indicated in the specification documentation. Increasing α to 35° brings these figures into agreement with the turning circle and the steering wheel turns.

Usually, the point of intersection of the steering arm angles is set to about 66% of the wheelbase at α = 15 - 25°. The error curve is flattened with increasing length of steering arm. In most cases, the steering arm length is made about 10-15% of the pivot centre distance P. In this case, the steering arm length is about 9.4% of the pivot centre distance.

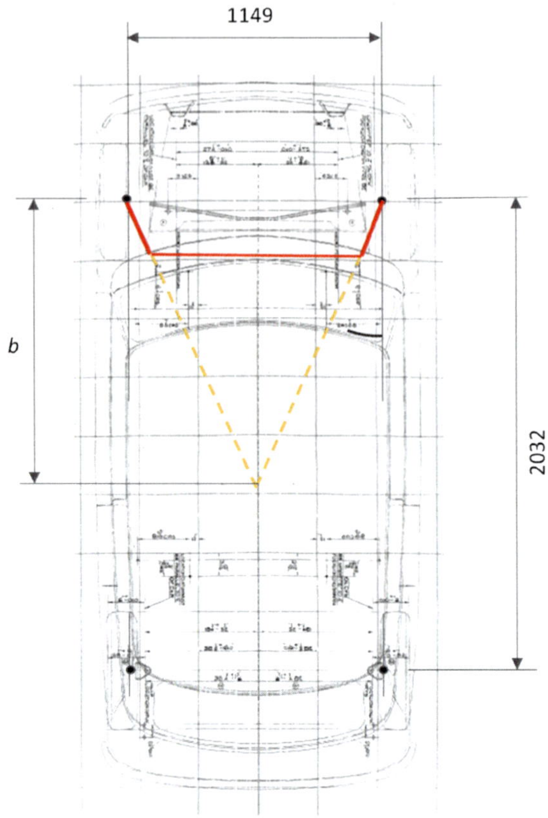

Fig. 8.1.10 *Practical steering arm angle intersection point, Morris Mini.*

Apart from the complication of the king pin inclination, the above discussion ignores several practical issues that also come into play such as castor and camber, and the effect of weight shift to the outer wheel under speed. However, a purely geometrical analysis does reveal some of the underlying issues that illuminate for us the basic principles of steering. The results shown here demonstrate the basis behind the excellent handling of this model that has evidently been the subject of considerable engineering design by BMC engineers.

8.1.7 Perfect Turning Austin 1800

A similar analysis can be performed for the Austin 1800 series with factory data shown in Table 8.1.2.

Wheelbase (W)	2694	mm
Track (T)	1422	mm
Camber	1.5	deg
Castor	1	deg
KPI	12	deg
Toe-in (straight ahead)	3.2	mm
Toe-out on turn (full lock)	23 ±1.5°(α), 20°(β)	deg
Tyre Width	150	mm
Tie rod length (t)	245	mm

Rack length (R)	779.4 (753.8)	mm
Lock to lock	3.8	Turns
Rack Travel (left or right)	83	mm
Turning Circle	11.3	m

Table 8.1.2 *Steering and suspension geometry ADO17.*

The results shown in 8.1.11 show that the rack and pinion analysis is not so favourable compared to the Mini, in that the range of angles over which perfect turning is achieved is less. The steering arms went through four revisions for this model during its lifetime with changes to castor as well which seems to indicate that the engineers were continuously attempting to optimise the geometry.

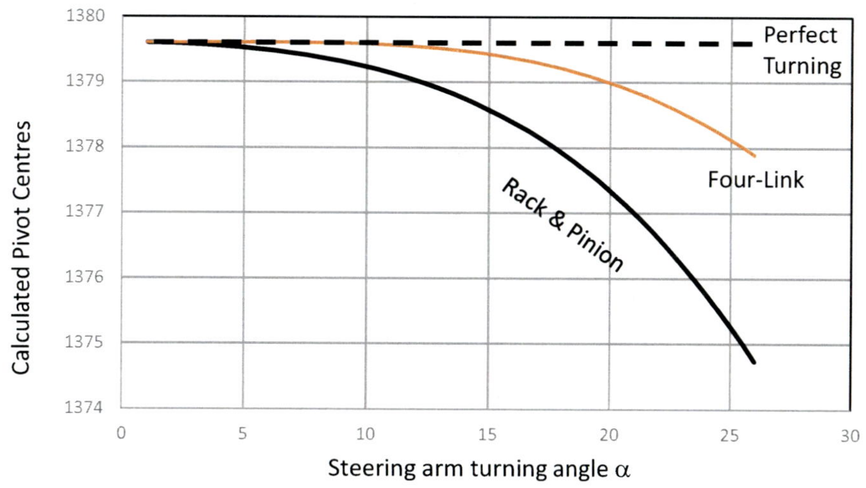

Fig. 8.1.11 *Error curve showing points at which perfect turning occurs ADO17.*

For this vehicle, the intersection of the steering arm angles occurs at 55% of the wheelbase measured from the front axle position.

CHAPTER 8. STEERING 169

CHAPTER 9. BRAKES

9.1 The Theory of Braking

9.1.1 Introduction

It is a surprise for most to learn that despite all the paraphernalia attached to the power unit – carburetters, exhaust, ignition, cooling system etc, and the attention given to torque and power output specifications, the brakes are more powerful than the engine. Occupying a relatively small amount of space, they are called upon to dissipate the kinetic energy of the vehicle without fuss, and in repeated applications. Their capability is familiar to anyone who compares the feeling of being thrown forward during heavy braking to that pushing rearwards under acceleration.

As shown in Chapter 2, the engine provides torque to the driving wheels, and a tractive effort is generated at the tyre contact path which propels the car forward. Braking is the reverse, in the sense that a braking tractive effort is generated at the contact patch opposite in direction to the tractive effort at the driving wheels during constant velocity or acceleration.

It is the function of the brakes to transfer the kinetic energy of motion of the vehicle to heat. The handbrake, which is often termed an emergency brake, is usually used to stop the vehicle rolling away when parked. It is the footbrake that is the focus of our attention here.

9.1.2 Load Distribution

Under braking, there is significant shift in weight to the front wheels as can be imagined, but precisely why this is so is deserving of some careful scrutiny.

The redistribution of load under braking has particular importance, not only in terms of its effect on handling, but also on the loads applied to suspension ball joints and other structural members.

At the static condition, the reaction force R_f front and rear (i.e. total reaction force for two wheels) is shown in Fig. 9.1.1. The new reaction forces under a deceleration a' are required, where a' is given by:

$$a' = \frac{a}{g}$$

That is, a' is the linear acceleration a in m/s^2 divided by g = 9.81 m/s^2. The new reaction forces are found by taking moments about a convenient centre. In fig. 9.1.1, the weight transfer q to the front wheels during forward braking is the desired quantity LD is the static load distribution over the wheels expressed as a ratio front to rear (e.g. 60/40).

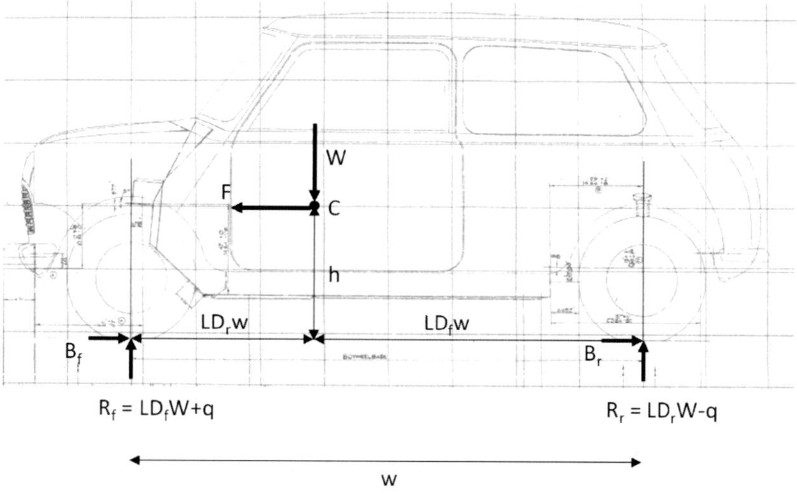

Fig. 9.1.1 *Braking and reaction forces.*

Taking moments about the front contact point shown in Fig. 9.1.1:

$$(LD_R W - q)w + Fh = W(LD_R w)$$
$$LD_R mg - q + ma\frac{h}{w} = LD_R mg$$
$$q = mg\frac{h}{w}a'$$

where m is the mass of the vehicle in kg and w is the wheelbase. This equation shows the weight transfer is proportional to the vehicle weight, the rate of deceleration a' (expressed as a multiple of g), and the ratio of the height of the centre of mass to the wheelbase.

The reaction forces R and weight transfers q shown above apply to the total of both wheels. Half of these values represent the forces at each wheel.

The front suspension and associated structures (e.g. tie road ball joints, lower arm mountings, engine mounts, etc) will be now called upon to support this increase in weight and so must be designed accordingly. When braking in reverse, similar calculations need to be made for the rear suspension components.

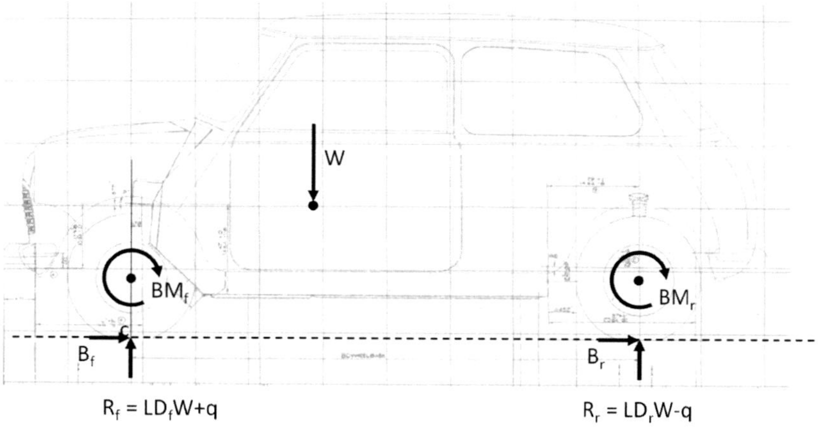

Fig. 9.1.2 *Application of brakes.*

Now, when the brakes are applied, the tractive effort force disappears from the driving wheels and is replaced by braking forces at the front and rear contact patch in the opposite direction on all four wheels. The vehicle, by virtue of is inertia, tends to keep the wheels rotating forward, while the brakes apply a moment BM at each wheel to slow them down.

The braking moment is provided by a friction force within the brake mechanism and is usually of the form of pads of some kind being forced into contact with a disc or drum, the force being controlled by the driver via the brake pedal.

Now, since the vehicle as a whole is decelerating a', then the total braking force $F = ma'g$. Ideally, the braking force on the front wheels should be in proportion to the weight over the front wheels, and that of the rear wheels to be in proportion to the weight over the rear wheels so that the limit to braking for front and rear would be reached at the same time.

The total braking force for an acceleration a', supplied by all four wheels is expressed:

$$F = ma'g$$

The sum of the normal forces equals the weight of the vehicle, and so:

$$\begin{aligned} W &= Rf + Rr \\ &= mg \end{aligned}$$

B_f and B_r added together equal the total braking force, and so:

$$\begin{aligned} F &= ma'g = B_f + B_r \\ &= Wa' \\ (Rf + Rr)a' &= B_f + B_r \end{aligned}$$

Since, under the transfer of weight under braking, R_f increases by q, and R_r decreases by q, then B_f and B_r should increase and decrease the same amount, so the front and rear braking forces should be:

$$\begin{aligned} B_f &= a'R_f \\ B_r &= a'R_r \end{aligned}$$

Note that *maximum* adhesion between the road and tyre occurs when the coefficient of friction is equal to the deceleration a'.

$$\begin{aligned} B_f &= \mu R_f \\ B_r &= \mu R_r \end{aligned}$$

For ADO15, Morris 850, factory specifications show that the centre of mass is h = 500 mm from ground level, and in line with the jacking points on the side sills. The wheelbase w is 2032 mm. The static weight distribution is 60/40 front to rear. The kerb mass is 650 kg.

Table 9.1.1 shows the weight transfer to the front and rear wheels, along with the ideal braking forces to be applied by the front and rear wheel brakes and also the maximum braking force[24] available when the coefficient of friction between the tyre and the road of $\mu = 0.6$.

[24] A friction force F reaches a maximum, i.e. the point of sliding, when $F = \mu N$ where μ is the coefficient of static friction.

Decelration	Weight Transfer	Wheel Loads R		Braking Force B (N)		Maximum Braking Force (N)	
a'	N	Front	Rear	Front	Rear	Front	Rear
0	0.0	3825.9	2550.6	0.0	0.0	2295.5	1530.4
0.1	156.9	3982.8	2393.7	398.3	239.4	2389.7	1436.2
0.2	313.8	4139.7	2236.8	827.9	447.4	2483.8	1342.1
0.3	470.7	4296.6	2079.9	1289.0	624.0	2578.0	1247.9
0.4	627.6	4453.5	1923.0	1781.4	769.2	2672.1	1153.8
0.5	784.5	4610.4	1766.1	2305.2	883.0	2766.2	1059.7
0.6	941.4	4767.3	1609.2	2860.4	965.5	2860.4	965.5
0.7	1098.3	4924.2	1452.3	3447.0	1016.6	2954.5	871.4

Table 9.1.1 *Computations of weight transfer, braking force, and maximum braking force for Morris Mini, ADO15.*

Note that the car, with perfect distribution of braking effort between front and rear wheels, cannot be brought to a stop without skidding at decelerations above 0.6*g*. The solid line in Fig. 9.1.3 shows the relationship between the front and rear wheel ideal braking forces at different values of deceleration.

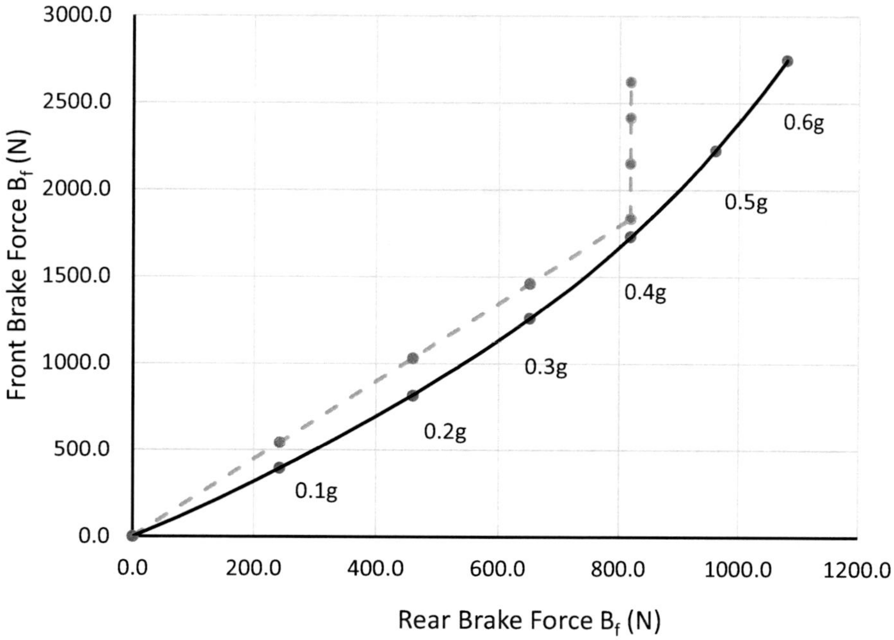

Fig. 9.1.3 *Front braking force vs rear braking force for increasing values of deceleration, Morris Mini ADO15. Dotted line shows computed actual distribution.*

The notable feature of Fig. 9.1.3 is that the ideal braking force split between front and rear wheels is non-linear. This then brings us to a consideration of just how the braking force is applied and what compromises are made to achieve close to an ideal balance. The dotted line shown in Fig. 9.1.3 is the actual brake distribution and will be discussed in Section 9.2.1.

In all the computations above, the reaction and braking forces are calculated for one side of the vehicle, i.e. one front wheel and one rear wheel. In reality, these forces are shared with those wheels on the other side so when determining loads on suspension components, the values of the forces involved (with the exception of the vehicle weight *w*) have to be halved.

9.1.3 Drum Brakes

Consider the drum brake system shown in Fig. 9.1.4. This is typical of the brake system to be found on the rear for many BMC/Leyland vehicles, and in some early models, the front also.

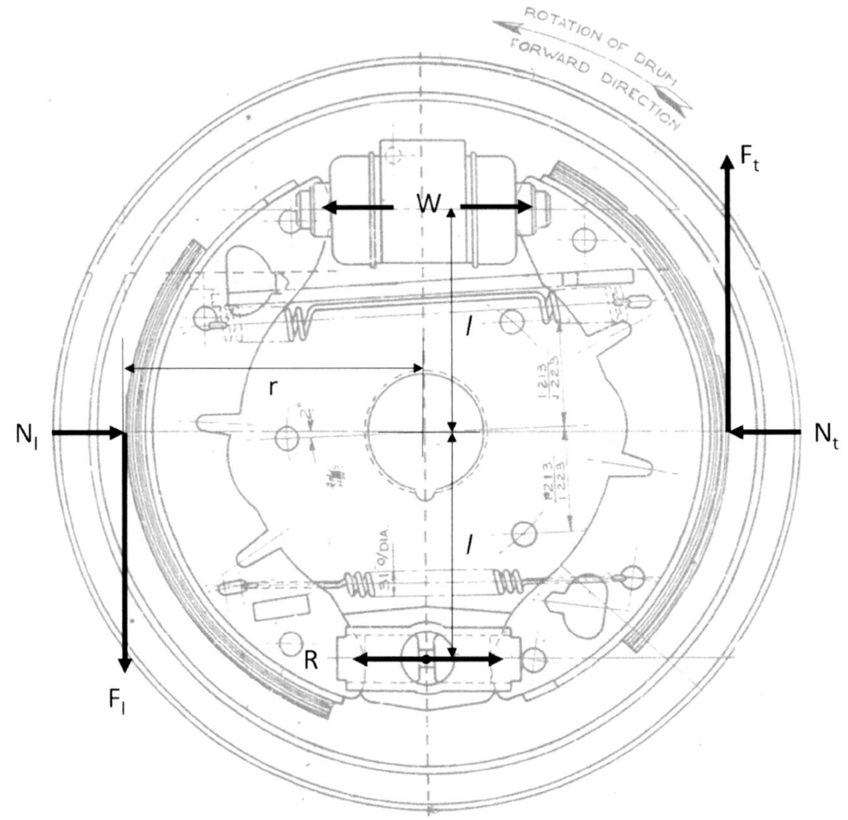

Fig. 9.1.4 *Drum assembly showing forward rotation for left hand side of the vehicle.*

When the pistons inside the wheel cylinder are pushed outwards under the action of hydraulic pressure, they apply an equal and opposite force on each brake shoe *W*. This in turn results in the show making contact with the drum (not shown in the picture), and the reaction normal force *N* is shown as it would act on each shoe. For simplicity, the forces can be thought of as acting at a single point opposite to each other as shown in Fig. 9.1.4.

The normal force thus leads to a friction force *F* acting on the shoe in the vertical directions. The magnitude of the normal force depends on the magnitude of the force *N* and the coefficient of friction μ of the contact between the lining and the inside drum surface. When there is relative movement of the shoe and drum, the friction force $F = \mu N$ where μ is the coefficient of sliding friction. It should not be assumed that the normal force on the left is the same as that on the right.

For the left hand shoe in Fig. 9.1.4, the friction force F_l creates a left hand turning moment about the pivot *P* since it acts perpendicularly through a distance *r*. The wheel cylinder force *W* also creates a left hand turning moment around the pivot acting through a distance 2*l*. The normal force N_l creates a right hand turning moment about the pivot *P* acting through a distance *l*. For mechanical equilibrium these moments must balance, and so:

$$\begin{aligned} 2lW &= N_l l - F_l r \\ &= N_l l - \mu N_l r \\ &= N_t(l - \mu r) \\ N_l &= \frac{2lW}{l - \mu r} \end{aligned}$$

The braking torque T_l, which eventually creates the braking force at the road surface, arises from the action of the friction force μN_l through distance *r*.

$$T_l = \mu N_l r$$

And so,

$$T_l = \mu r \frac{2lW}{l - \mu r}$$

For the right hand shoe, the friction moment F_t acts opposite to that generated by *W*, and in the same direction as that generated by N_t, and so by a similar analysis to the above, we obtain:

$$\begin{aligned} 2lW &= N_l l + F_t \\ &= \mu N_t r + N_t l \\ N_t &= \frac{2lW}{l + \mu r} \\ T_t &= \mu r \frac{2lW}{l + \mu r} \end{aligned}$$

Since the denominator for T_l is always smaller than that for T_t, then T_l will always be greater than T_t by an amount which depends on the friction coefficient, and the radii *l* and *r*. The shoe on the left in Fig. 9.1.4 is called the leading shoe, and that on the right, the trailing shoe. Thus, in the case of the leading shoe, the friction force moment acts with the normal force moment, whereas with the trailing shoe, it acts against it.

The situation is reversed when braking in reverse. In the Mini range, in early vehicles, both front and rear drum brakes have both leading and trailing shoes with a single, double acting wheel cylinder. In later models, the drum brakes on the front were two leading shoes (for forward motion) with two single acting wheel cylinders.

For the Morris 850, ADO15, the dimensions for *r* and *l* are 88.3 mm and 65 mm respectively. Fig. 9.1.5 shows the ratio of T_l/T_t for various values of the shoe to drum friction coefficient. A representative value of shoe to drum friction coefficient is 0.3.

Fig. 9.1.5 shows that the contribution of the leading shoe to the overall braking torque increases with increasing friction coefficient. At $\mu = 0.3$, the contribution is almost double. It can therefore be expected that the leading shoe lining material will wear more quickly than that of the trailing shoe.

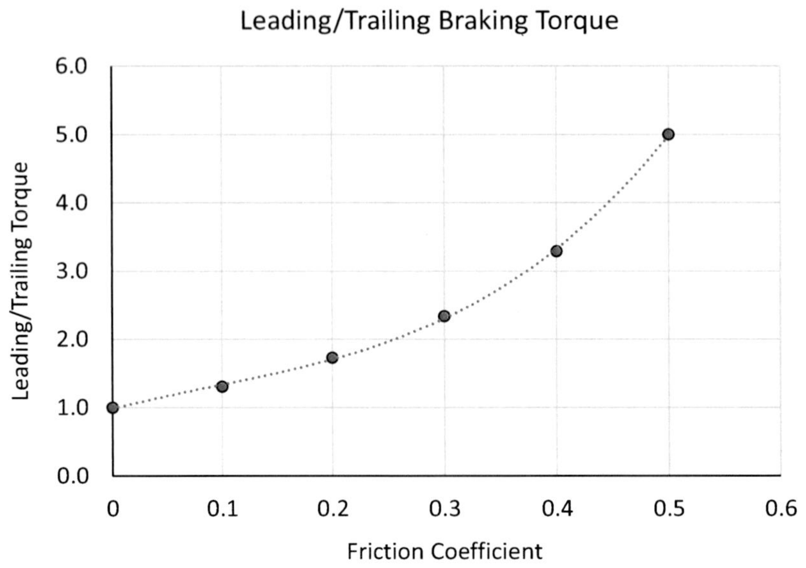

Fig. 9.1.5 *Ratio of braking torque for leading shoe and trailing shoe vs friction coefficient at W = 500N.*

Consequently, calculations show that for a leading and trailing shoe combination, more force is needed at *W* to produce the same braking torque compared to a twin leading shoe arrangement. The significance of this is apparent when the sizes of the wheel cylinders for the two different systems are compared.

The braking torque, or moment, generated by the brakes produces a braking force at the tyre contact patch with the road, the moment arm being the rolling radius of the wheel and tyre. Skidding occurs when the maximum friction force equals the normal force times the coefficient of static friction between the tyre and the road – the latter usually taken to about about 0.6.

A similar analysis can be made for disc brakes where it is seen that for a particular braking torque, a greater force *W* is required to be applied perpendicular to the discs due to the smaller radius at which the disc is to the hub compared to an equivalent drum brake, and the lack of "servo" action of the leading brake shoes. Consequently, vehicles fitted with front disc brakes usually are accompanied by a vacuum assisted booster that provides an increase in hydraulic pressure for the front brakes only.

The effects of weight transfer under braking have particular significance when a vehicle is towing a trailer. If the trailer wheels are unbraked, then the inertia forces of the trailer generated by the vehicle brakes are transmitted through the tow bar and affect the weight transfer to the front and rear suspension. If the trailer wheels are braked electrically or mechanically, then there is a weight transfer downward on the tow hitch that has to be accounted for in the design of the tow bar and the draw bar.

9.1.4 Master and Slave Cylinders

The force at the brake show imposed by the wheel cylinder is generated from hydraulic pressure. The driver applies a force to the brake pedal, which acts on the master cylinder piston producing pressure in the brake lines. The force imposed on the wheel cylinder at this pressure is:

$$W = pA$$

Where *A* is the area of the wheel cylinder, and *p* is the hydraulic pressure at the wheel cylinder.

Previously, it was shown that to account for weight redistribution under braking, the braking force at the front wheels is required to be greater than that of the rear wheels. Consequently, if the general dimensions of the brakes front and rear are the same, then the force *W* at the wheel cylinders for the front brakes has to be greater than that for the rear brakes. Thus, if the hydraulic pressure is the same front and rear, the area, and hence diameter, of the front wheel cylinders is required to be greater than that of the rear.

But, how much greater? The amount depends non-linearly on the deceleration, and, as shown above, depends also on whether the brakes are of the leading and trailing design or twin leading. Evidently, some compromises are needed to provide adequate braking under all anticipated conditions.

9.1.5 Braking Power

Apart from preventing the vehicle from rolling away when stopped, it is the function of the brakes to convert kinetic energy of motion into heat. In this sense, brakes are 100% efficient because heat is the desired outcome. By contrast, most physical processes involve undesirable heat generation which leads to inefficiencies.

Heat is produced by friction at the interface of the brake linings and the brake drum (or disc pads and disc). Heat is dissipated via radiation from the drum, conduction from the drum to the wheel bub and wheel, and conduction from the brake shoes to the back plate. Heat may also be dissipated via convection or conduction via the air flow around the brakes.

The coefficient of friction at the brake shoe is affected by temperature. If the heat generated by the brakes cannot be dissipated at the same rate at which it is generated, the temperature of the shoe and brake drum will rise. If the temperature rises too much, the coefficient of friction may reduce to dangerous levels – a condition known as brake fade. As well, excess temperatures at the wheel cylinder may cause the hydraulic fluid to boil and consequent increase in pedal travel to the extent that the pedal strikes the floor and no more braking force can be applied.

The total area of the linings that is the important parameter as far as power dissipation is concerned. A wide brake shoe does not offer any additional braking effort over a narrow one. Rather, a wide brake shoe increases the dissipative power of the brakes, and the lining may wear at a lower rate compared to a narrow shoe if that power is not always utilised.

In the introductory remarks to this Chapter, it was said that the brakes are more powerful than the engine. To illustrate this, consider a 650 kg motor vehicle travelling at 100 km/hr (27.8 m/s) that is brought to rest in a distance of say 50 m by the application of the brakes.

The kinetic energy to be dissipated is:

$$\begin{aligned} KE &= \frac{1}{2}mv^2 \\ &= 0.5(650)(27.8)^2 \\ &= 251kJ \end{aligned}$$

The deceleration can be found from:

$$\begin{aligned} v^2 &= u^2 + 2ad \\ 0 &= 27.8^2 + 2a(50) \\ a &= -7.7m/s^2 \end{aligned}$$

where u is the initial velocity and $v = 0$ is the final velocity.

The time taken is thus:

$$\begin{aligned} v &= u + at \\ t &= \frac{27.8}{7.7} \\ &= 3.6s \end{aligned}$$

And so the power is:

$$\begin{aligned} P &= \frac{KE}{t} \\ &= \frac{251}{3.6} \\ &= 69kW \end{aligned}$$

In Chapter 2, it was found that at maximum acceleration through the gears, it took a 37 kW 1098cc engine about 20 seconds for this vehicle to reach 100 km/h from a standing start – overcoming air and rolling resistances, and other losses in the mechanical components in the process.

The brakes can bring the vehicle to rest from that speed in about 3.5 seconds[25], dissipating energy at the rate of about 70 kW in the process at 100% efficiency. The brakes occupy little space, do not require a water cooling system, exhaust system, fuel system, and ignition system, and make little noise. Their only input is a moderate push on the pedal on the part of the driver.

9.1.6 Stopping Distance

Of practical interest to both motorists and engineers is the stopping distance from a particular vehicle speed. The stopping distance is the sum of the distance travelled during the driver reaction time (taken to be a constant 0.75 seconds), and the braking distance (i.e. that covered by the vehicle while brakes are applied).

The braking distance can be readily computed in terms of the deceleration from kinematics equation:

$$\begin{aligned} v^2 &= u^2 + 2ad \\ d &= \frac{v^2}{2a} \end{aligned}$$

Expressed in terms of deceleration as a fraction of g, and in km/h, we have the braking distance d in m as:

$$d = \frac{(km/h)^2}{254.275a}$$

The notable feature is that the braking distance depends upon the square of the speed. The distance travelled during the driver reaction time depends linearly upon the speed. Fig. 9.1.7 shows how stopping distance varies as a function of vehicle speed and includes driver reaction time.

---

[25] Wheels Magazine July 1961 quote 2.7 seconds braking from 96 km/h for Morris 850.

178 THE BMC ENGINEERING COMPANION

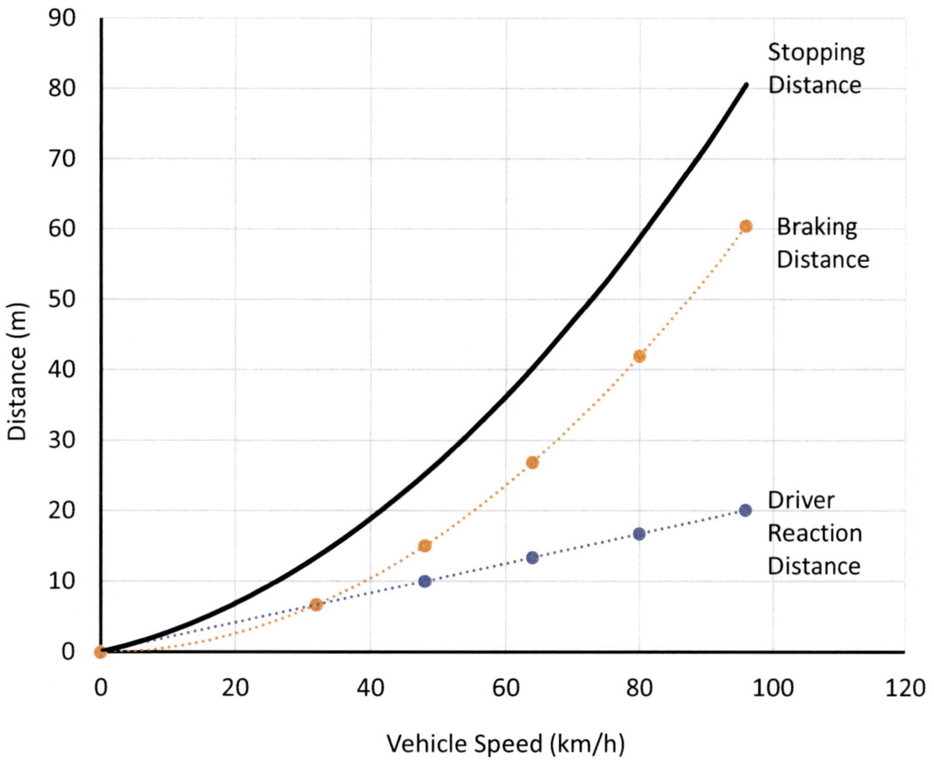

Fig. 9.1.7 *Driver reaction, braking and stopping distances vs vehicle speed for a deceleration of 0.6g.*

As an example, a road test[26] on a Morris 850 reports a stopping distance of 8.5 m from 48 km/h, which according to the above formula, represents a deceleration of 1g.

A comparison of the equations above for stopping time and stopping distance reveals that the time taken to come to a stop from some initial speed depends linearly on that speed, whereas the distance covered to come to a stop depends on the square of the initial speed. Whether or not a particular vehicle can stop within the distances shown depends on the capability of the brakes to provide the requisite deceleration, which in turn depends upon the brake system design and the mass of the vehicle.

9.2 Morris 850 ADO15

9.2.1 Foot brake

Fig. 9.2.1 shows the brake pedal and details of the master and slave cylinders for a Morris 850 ADO15 (approx. mid 1963 specification) with leading and trailing show brakes at the front and rear with a 0.7" diameter master cylinder, 15/16" diameter wheel cylinders at the front, and 5/8" diameter wheel cylinders at the rear.

The 7" (178 mm) diameter brake drums are operated on by shoes of size 6.75" length by 1.24" wide, and have an area per wheel of 16.88 sq ins. The rolling radius for each wheel is 252 mm.

[26] Modern Motor magazine May 1961

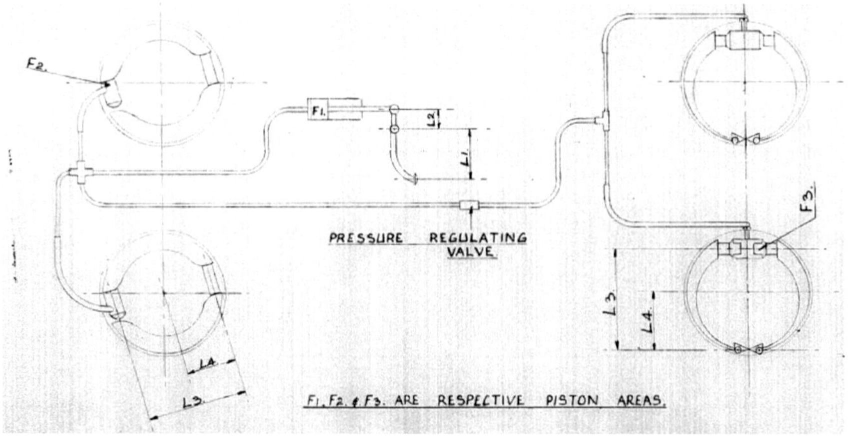

Fig. 9.2.1 *Brake leverage ratio, Morris Mini, ADO15 (674).*

The brake pedal has a leverage ratio of 4.2. At the wheel, the wheel cylinder has a mechanical leverage ratio 1.93 (wheel cylinder with respect to the bottom pivot). There is a hydraulic ratio of 1.79 from master cylinder with respect to the wheel cylinder area, and so the overall force leverage for the front brakes is:

$$\begin{aligned} LR_f &= 4.2(1.93)(1.79) \\ &= 14.4 \end{aligned}$$

This is the leverage ratio with the wheels immobile (i.e., the vehicle stationary with a force applied to the brake pedal and so no friction force) and the force generated at the centre of each of the four front brake shoes.

For the rear wheels, the leverage ratio is:

$$\begin{aligned} LR_r &= 4.2(1.93)(0.796) \\ &= 6.415 \end{aligned}$$

Consider the input data shown in Table 9.2.1:

Quantity	Value	Unit
Dimension l	0.075	m
Dimension r	0.1	m
Shoe friction coefficient	0.3	
Foot pedal	245	N
Rolling radius	0.225	M
Vehicle mass	650	kg
Master cylinder piston dia	0.7	in
Front wheel cylinder piston dia	15/16	in
Rear wheel cylinder piston dia	5/8	in

Table 9.2.1 *Input data for brake force calculations.*

Using the dimensions shown in Fig. 9.2.1, and the data in Table 9.2.1, the following quantities are thus calculated:

Master cylinder force	1020.8N
Master cylinder pressure	4.110MPa
Wheel cylinder W force (front)	1.8 kN
Wheel cylinder W force (rear)	814.0 N

One wheel		
Front brake torque T_l	240.1	Nm
Front brake torque T_t	87.3	Nm
Rear brake torque T_l	106.8	Nm
Rear brake torque T_t	38.9	Nm
Front Road Force Leading	1067.2	N
Front Road Force Trailing	388.1	N
Rear Road Force Leading	474.8	N
Rear Road Force Trailing	172.7	N
All wheels		
Total (Front) 4 wheels	2910	N
Total (Rear) 4 wheels	1295	N
Total braking force	4205.7	N
Deceleration	6.47	m/s2
	0.66	g
Front to rear brake ratio	2.24	

Table 9.2.2 *Results of computations for braking force for each shoe, and deceleration of the vehicle for 25 kg foot pedal effort.*

The results shown in Table 9.2.2 are consistent with those carried out in other Sections. No account of the effect of the brake limiter that is fitted to this vehicle has been included.

The front to rear braking effort is computed as 2.24 and this is indicated by the dotted line in Fig. 9.1.3. It appears therefore that the front to rear braking distribution has been set up for almost perfect braking at a deceleration of approximately 0.45*g*, moderate to heavy braking.

9.2.2 Handbrake

The handbrake in most vehicles is usually operated mechanically rather than hydraulically so that it may have a better chance of acting as an emergency brake in the event of a hydraulic failure. In modern times, failure of the hydraulic brakes is a rare event, and so the emergency brake is more often regarded as a parking brake.

The leverage ratio for the handbrake for a Morris Mini, is shown in Fig. 9.2.1 and is given as 20.67:1.

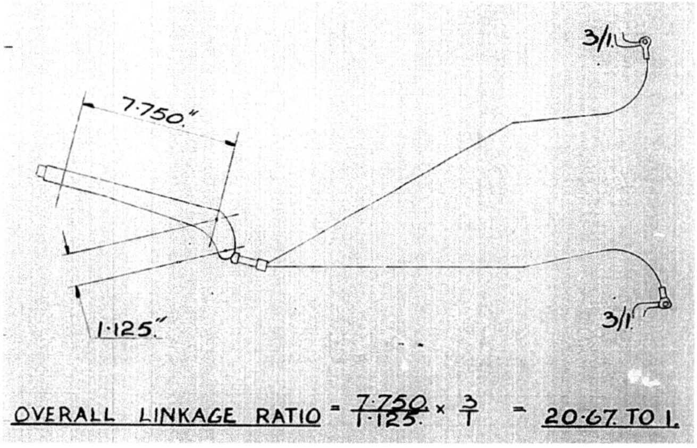

Fig. 9.2.2 *Handbrake leverage ratio Morris Mini, ADO15 (674).*

9.2.3 Rear Brake Pressure Limiting Valve

In the Morris Mini range, there is a pressure limiter valve 21A2031 (370 psi, later increased to 390 psi) in the line to the rear brakes. This limits the pressure applied to the rear brake drums so as to avoid lockup as weight is transferred to the front.

As an example of the pressures involved, 25 kg applied by foot on the brake pedal will result in a hydraulic pressure of about 4 MPa (600 psi) in the brake lines, limited to 370 psi (2.7 MPa) for the rear brakes.

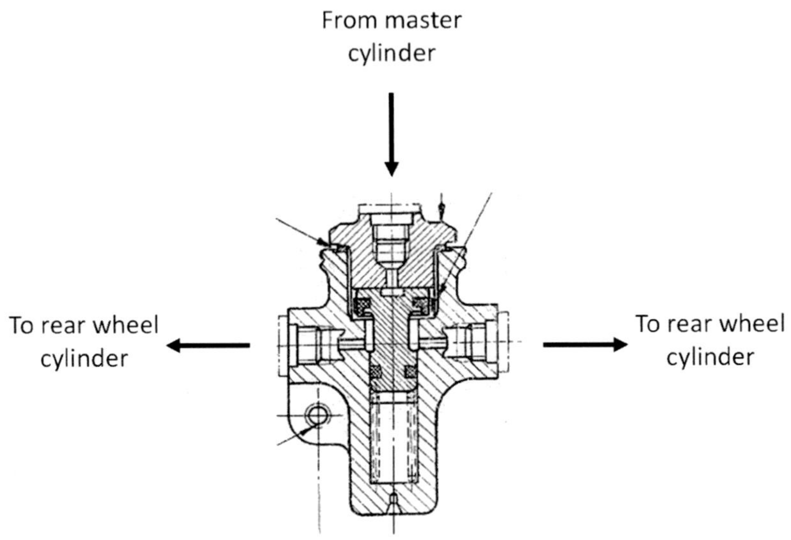

Fig. 9.2.3 *Section view 21A2031.*

The effect of this is to limit the rear braking force to about 800 N (for $\mu = 0.6$ at road) and this is shown as the vertical dotted line in Fig. 9.1.3.

CHAPTER 10. ELECTRICAL

10.1 Voltage and Current

Experiments show that, for a particular piece of material, when the applied voltage *V* is increased, the current *I* increases in the same proportion. In other words,

$$V = IR$$

The constant of proportionality is called resistance, *R*. This is Ohm's law.

When a voltage is applied to a conductor such as a copper wire, the mobile electrons within the wire are accelerated by the electric field, then collide with other electrons and atoms losing kinetic energy as electromagnetic radiation (i.e. heat). Eventually the net average velocity of the electrons reaches a steady value called the drift current, and the rate of flow of charge through the wire is constant. This is direct current, DC.

If the applied electric field, i.e. the voltage, is periodically reversed in polarity, the current changes direction periodically as well. This is alternating current, AC.

The rate of conversion of electric energy into heat is the power *P* dissipated by the circuit and is calculated from both the voltage *V* and the current *I*, of the current I and the resistance *R*.

$$\begin{aligned} P &= VI \\ &= I^2R \end{aligned}$$

Significant complications arise when components like capacitors and inductors are present in a circuit since the resistance to the flow of current in these devices depends on the rate of application of voltage and current to them (Chapter 4).

What is not often appreciated are the difficulties arising from inductances. An inductor is a coil of wire which, when current passes through it, forms an electromagnet. However, when the current is increasing, there is a backwards acting voltage applied to resist the incoming current. When the current is turned off, the magnetic field in the inductor reduces and the changing field induces a voltage in the coil that attempts to keep the current running. Both these effects play a significant role in the operation of the ignition coil.

10.2 Battery

A battery is a collection of electrochemical cells arranged in series. In a motor vehicle lead acid battery, there are six cells, each producing about 2V, arranged in series to provide 12V for the vehicle. The oxidation-reduction chemical species involved are:

$$PbSO_4(s)/Pb(s),SO_4^{2-}$$
$$E^o = -0.36V$$

and

$$PbO_2(s),H^+,SO_4^{2-}/PbSO_4(s)$$
$$E^o = +1.69V$$

The quantity E^o what is called the standard potential for that species and determines in which direction reactions occur. In this case, since the standard potential for the PbO_2 reaction is higher, then this species undergoes reduction and is the cathode (the positive terminal of the battery).

$$PbO_2(s)+4H^+ +SO_4^{2-}+2e^- \rightarrow PbSO_4(s)+2H_2O$$

The other reaction proceeds at the anode, the negative terminal.

$$Pb(s)+SO_4^{2-} \rightarrow PbSO_4(s)+2e^-$$

When discharging, the overall chemical reaction is:

$$PbO_2(s)+4H^+ +2SO_4^{2-} \rightarrow 2PbSO_4(s)+2H_2O$$

The concentration of sulphuric acid H_2SO_4 in a motor vehicle battery is usually about 5M and so the actual cell potential is about 2.15V. The anode (the negative plates) is solid lead, Pb, while the cathode (the positive plates) is lead peroxide, PbO_2. The reaction product, lead sulphate $PbSO_4$, sometimes appears as a white precipitate on the battery terminals.

As the cell discharges, the concentration of H_2SO_4 decreases and the density of the solution also decreases. The density of the solution is thus a measure of the state of charge of the battery and is measured with a hydrometer.

When the battery is charged, the above reaction proceeds in reverse. Overcharging results in the evolution of gaseous H_2 at the anode and O_2 at the cathode, and the water in the electrolyte has to be replenished.

The operation of the battery is conveniently through of in terms of a source of electric potential connected in series with an internal resistor.

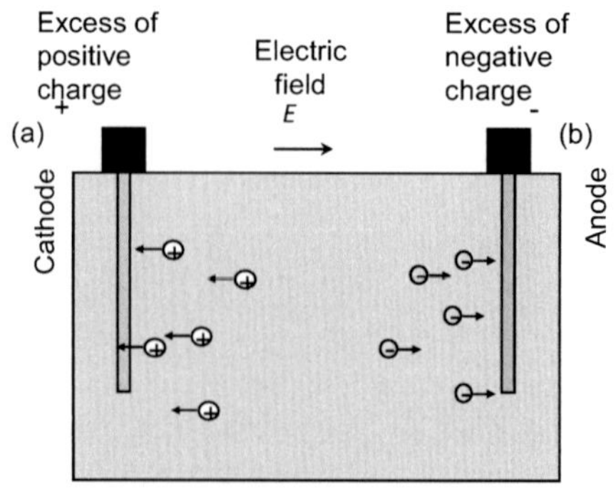

Fig. 10.2.1 *Electrochemical cell, open circuit.*

Consider an electrochemical cell. Oxidation-reduction reactions cause attractions between ions at the electrodes such that positive electrical charges to build up at (a) and negative charges at (b). Electrostatic repulsion due to build-up of positive charge at (a) eventually becomes equal to the chemical attractions tending to deposit more positive charges and the system reaches equilibrium. The emf (electromotive force) is defined as the amount of energy expended by the cell in moving 1 coulomb of charge from (b) to (a) *within the cell*. At open-circuit, emf = V_{ab}. This is called the open-circuit terminal voltage.

Now is an external load R_L is connected across (a) and (b). The terminal voltage V_{ab} is now reduced. Loss of positive charge from (a) reduces the accumulated charge at (a) and hence chemical reactions proceed and more positive charges are shifted from (b) to (a) within the cell to make up for those leaving through the external circuit. Thus, there is a steady flow of positive charge through the cell and through the wire. The "wires" in Fig. 10.2 are drawn to accentuate the voltage drops in the circuit.

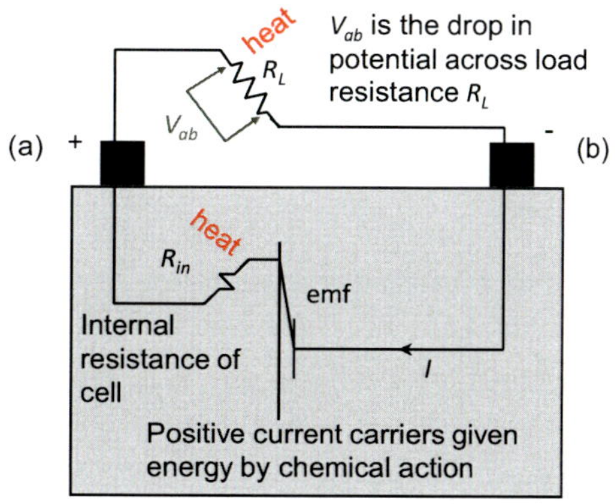

Fig. 10.2.2 *Electrochemical cell with current draw.*

But, the continuous conversion of chemical potential energy to electrical energy is not 100% efficient. It is as if charge moving within the cell encounters internal resistance, which, in the presence of a current I, means a voltage drop so that at closed-circuit, emf = V_{ab}+ IR_{in}.

The concept of an internal resistance is extremely important and aids in both the analysis of electrical circuits, and also provides a physical understanding of what is happening. In circuit analysis, Thevenin's theorem allows us to replace a complex circuit by an equivalent circuit containing a source of emf V_T and an internal resistance R_{int}.

To see how this works, consider the following circuits.

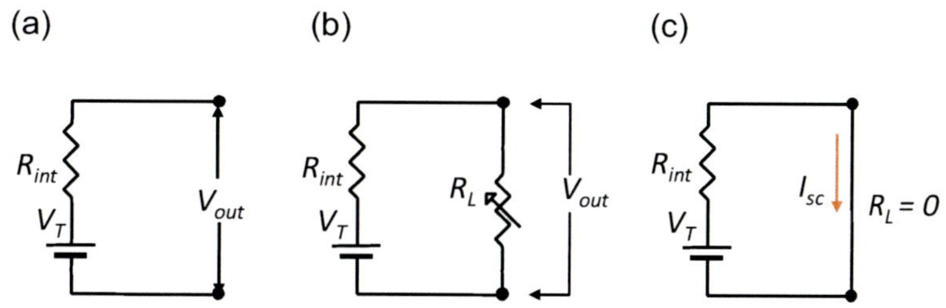

Fig. 10.2.3 *Thevenin equivalent circuit.*

In (a), with no load connected across the output, the output voltage $V_{out} = V_T$. This is the open-circuit voltage $V_{oc} = V_T$. In effect, the "load resistance" is infinite. In (b), if we reduce the load resistance R_L from an infinitely high value, the current drawn from the power supply increases and the output voltage V_{out} decreases. Why? Because of the voltage drop across R_{int}. As $R_L \rightarrow 0$, the current reaches a maximum called the short-circuit current and all of V_T is dropped across R_{int}. With $R_L = 0$, it is evident that all of V_T must be dropped across R_{int}.

That is, in practice, V_T and I_{sc} can be measured by measuring the open-circuit voltage and short circuit current. It is strongly advised that the short circuit current of a motor vehicle battery not be measured directly since the current is extremely large, and the resulting heat dissipated in the "internal resistance" would most likely cause an explosion. However, the internal resistance can be estimated from the voltage drop at the battery terminals when the starter motor is operated and the current through the starter motor is measured. Some early design battery testing machines impose a heavy current drain on the battery and monitor the terminal voltage as a measure of battery health. For a 12V motor vehicle battery, the internal resistance is a few mΩ.

V_T and R_{int} are useful tools for reducing a complicated power supply circuit to a simpler circuit. This is Thevenin's theorem. That is, any two-terminal voltage source, no matter how complicated, can be represented by V_T and R_{int}.

If a battery is through of as containing a cell and an internal resistance, then a much better understanding of what happens in a vehicle electrical system can be had. For example, when a high current device is activated, such as the starter motor, it is observed that the headlamps dim. This is because even though the interior cell voltage may remain the same at 12.15V, the voltage drop across the internal resistance increases as more current is drawn from the battery, and consequently, the terminal voltage decreases and the headlamps dim.

The internal resistance increases as the cell deteriorates with age and use to the point where it cannot be reversed by charging and so little if no current can be drawn before all the available cell voltage is dropped across it. The battery is flat.

10.3 Starter Motor and Generator

10.3.1 Electromagnetic Induction

The operation of the starter motor and the generator relies on the remarkable circumstance that if a charged particle, whether positive or negative, with a charge *q* moves with a velocity *v* in a magnetic field of strength *B*, a force *F* is exerted on that particle. In circuit analysis, the charged particle that moves is taken to be positively charged for historical reasons.

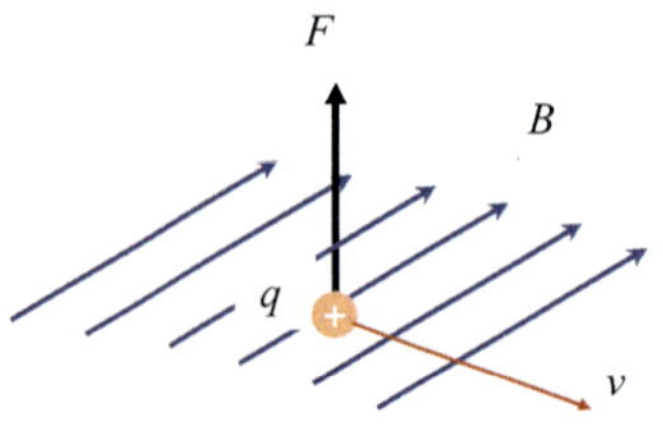

Fig. 10.3.1 *Force experienced by a charged particle in a magnetic field.*

In fact, the charged particles that actually move in a copper wire are the negatively charged electrons, but the equations come out exactly the same no matter which way it is described.

If the moving charges are of the form of an electric current in a wire, then the force experienced by the wire is given by:

$$F = ILB$$

Where I is the current in amps, L is the length of the wire, and B is the strength of the magnetic field in Tesla. If the wire forms a loop, and can revolve, then an electric motor is made which can then produce mechanical work. This then is the basis for a starter motor in a motor vehicle.

The opposite is also true. If the wire is now moved in the magnetic field mechanically, then the force so induced acts on the mobile charges in the wire (i.e. the electrons), and a current is produced. This is the basis for the operation of a generator in the charging system of a motor vehicle. For both electric motors and generators, it is the perpendicular component of the particles direction of motion to the magnetic field that produces the effect.

In Fig. 10.3.2, the production of an electric current can be visualised by the way in which the conductor is placed in the magnetic field, and the consequences of its rotation.

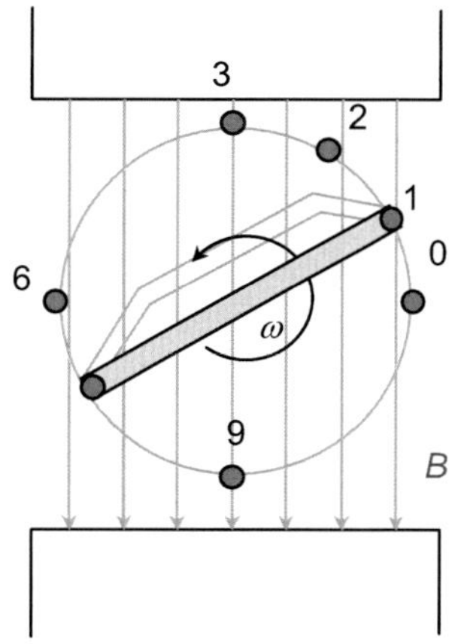

Fig. 10.3.2 *AC voltage and current generated by the rotation of a loop in a magnetic field.*

- at (0) motion of conductor is parallel to B, hence induced voltage = 0
- at (1), conductor has begun to cut magnetic field lines B, hence some voltage is induced
- at (2), conductor cuts magnetic field lines at a greater rate than (1) and thus a greater voltage is induced
- at (3), conductor cuts magnetic field lines at maximum rate, thus maximum voltage is induced
- from (3) to (6), the rate of cutting becomes less
- at (6), conductor moves parallel to B and $v = 0$
- from (6) to (9), conductor begins to cut field lines again but in the opposite direction, hence induced voltage is reversed in polarity.

In a motor vehicle generator, several loops are wound on to the armature (the rotating inner) and their ends are brought out to copper terminals at the commutator. Each end of one of the coils is momentarily connect to brushes on opposite sides of the commutator and so the output becomes a somewhat rippled DC as each coil is brought into operation with the rotation. This is a mechanical form of rectification. In an alternator, the AC output is converted to DC using a bridge circuit with semiconductor diodes.

In generator operation, magnitude of the voltage and the current so produced depends on the angular velocity ω. In motor operation, the resulting angular velocity depends on the voltage, and hence current, supplied to the coils.

Despite their similarities in principle of operation, the starter motor and generator are operated in a motor vehicle under quite different circumstances. The starter motor operates for a few seconds only, draws and heavy current of some 200 to 300A, and supplies a fairly large torque to the armature shaft. In contrast, the generator operates at high speed, for long periods, and the current output is about 20 or 30A or so. It is not surprising therefore to find that the nature of the brush and commutators in each are designed to account for their operational requirements.

10.3.2 Generator Output Regulation

Early model BMC vehicles are fitted with a Lucas RB106/2 regulator which consists of two components, a cut-out and a voltage regulator. When viewed from the front, the right-hand side coil assembly is the cut-out while that on the left had side is the voltage regulator.

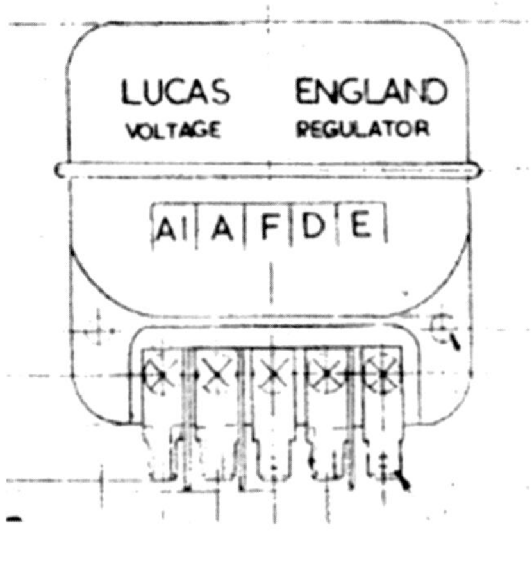

Fig. 10.3.3 *Lucas RB106/2 Regulator (13H203).*

It should be noted that the frame of the regulator unit is at generator output voltage potential.

The shunt winding on the regulator (fine wire) is fed by the generator output. When the output voltage reaches a certain point, the armature is pulled down and the connection between the output and the field coils is broken, thus interrupting the generator output. That is, the voltage output from the generator is controlled by controlling the current in the field coils. The opening and closing of the regulator contacts happens at about 60 times per second. The spring adjustment on the contacts controls the charging voltage which is rated at 15.4 – 16.4V at an engine speed of 1500 rpm for a typical BMC A series engine.

The cut-out bobbin disconnects the battery from the generator when the generator output voltage is lower than the battery voltage. This prevents the battery from discharging through the generator. Cut-in volts (when points close upon rising engine speed, thus connecting the rising generator output to the battery) is specified at between 12.7 – 13.3V.

The red ignition warning light in the instrument housing is fed with battery voltage on one side of the filament and generator output voltage (terminal D) on the other side. When the generator voltage drops below the battery voltage, current flows from the light through the generator armature windings to earth and the light glows indicating reduced charging from the generator.

In a two-bobbin arrangement, there is an indirect control over the charging current via the compensating series winding (as opposed to the shunt winding) on the regulator bobbin. This effectively lowers the voltage at which the regulator points are opened thus limiting the charging current. Terminal A is connected to the battery and passes through a heavy series winding in the regulator unit to terminal A1 which feeds the fuse box and then to the ignition switch and other heavy current circuits such as the lights. A heavy winding on the regulator bobbin between terminal A and A1 is a load sensing device that assists in preventing the generator from supplying too much current. Terminal D is the generator output and F connects to the field coils. The RB106 unit is thus referred to as a compensated voltage control device.

Terminal D is the generator output and F connects to the field coils.

The designation RB106 means "Regulator Box" and 106 indicates no internal fuse.

10.3.3 Alternator

An alternator has several advantages over a generator, the main one being the charging function occurs at a lower engine speed. As well, it is smaller and lighter than a generator, requires less maintenance and the brushes carry a smaller amount of current. Rectification of the AC signal is done via diodes.

The rotor in an alternator is essentially an electromagnet, and act as the "field" windings. The magnetic field rotates with the rotation of the rotor. The windings on the stator, (i.e., the housing) being stationary, are thus acted upon by a changing magnetic field and a current is then induced in them. That is, unlike a generator, the charging current is produced in the stationary windings.

The rectification function of the commutator in a generator is replaced by diodes in the alternator. A combination of diodes converts AC into DC. The diodes also prevent backflow of current and so no cutout relay is required.

Both the alternator and generator require voltage regulation to prevent over charging of the battery. However, an alternator is self-regulating with respect to current and so no external current regulation is required.

10.4 Starter Motors[27]

At the time of introduction in 1961, starter motors for most BMC vehicles were most likely UK supplied for CBU power units. However, within a few years this changed to local supply by Lucas Australia. For the Morris 850, there are three different types of starter motor listed in the Service Parts List.

Part No. 25071 – most likely a UK produced item. The following details are listed.

Model	25071 M35G
Origin	Most likely UK
Pinion	54250372
Bracket (drive end)	5425032 (cast iron)
Bush (drive end)	253316
Field Coils	54250820
Armature	54259822
Brushes	251108 or 62250270
Bracket (commutator end)	251935 or 54249301
Rear Bush	251477
Spring retainer	Nut

Table 10.4.1 *Part No. 25071.*

This is shown for Engine Number up to 8AM271755.

Listed as an alternative to the above is Part No. 25075.

Model	25075 M35G
Origin	Most likely UK
Pinion	250698
Bracket (drive end)	5425032 (cast iron)
Bush (drive end)	253316
Field Coils	251089
Armature	5420096
Brushes	251108 or 62250270
Bracket (commutator end)	251935 or 54246004 or 54259301
Rear Bush	25083
Spring retainer	Nut

Table 10.4.2 *Part No. 25075.*

[27] Acknowledgement: GB's Mini & Moke World, Brookvale

From 8AM271756 onwards, we have Part No. 25083.

Model	25083 M35G
Origin	Most likely UK
Pinion	54252212
Bracket (drive end)	54254175 (alloy)
Bush (drive end)	253316
Field Coils	251089
Armature	54254042
Brushes	251108 or 62250270
Bracket (commutator end)	251935
Rear Bush	251477
Spring retainer	Clip

Table 10.4.3 *Part No. 25083.*

Notice already that for 1961, there are three different starter motors with most important components like the field coils and armature not interchangeable, although they may all use the same brush set and front and rear bushes. The last one listed above has the circlip end for the main spring rather than the slotted nut and split pin.

Now, moving into 1962 and onwards, Part No. 62925008 is specified for all models "except Morris 850".

Model	62925008 M35G
Origin	Australia
Pinion	54252212
Bracket (drive end)	54254175 or 54257544 (alloy)
Bush (drive end)	253316
Field Coils	672011
Armature	54254042
Brushes	62250270
Bracket (commutator end)	54259301 or 251935
Rear Bush	251477
Spring retainer	Clip

Table 10.4.4 *Part No. 62925008.*

Part No. 62925008 appears to be an Australian version of Part No. 25083.

Despite the previous starter being specified for all models except Morris 850, for the Morris Mini Moke, there is Part No. 62925008 (same as above) but with a choice of end backets 54256004 or 54259301.

Next up is Part No. 62925048, similar to previous, except a choice of field coils 672011 or 62250637 or 251809.

All the above starter motors are type M35G, the 35 indicating a 3.5" yoke and the G indicating a cast end bracket at the commutator end.

UK models appear to have a cast iron front bracket (drive end) while Australian models tend to have an aluminium alloy front bracket, but not always. Early Australian starter motors, e.g. 62925009 (which can also be found) have a cast iron front bracket.

Even the cast end bracket can have a different appearance between early and later models.

Fig. 10.4.1 *End bracket differences.*

The M35G starter motors have a conventional commutator with copper segments along the axis of the armature as shown in Fig. 10.3.2:

Fig. 10.4.2 *Commutator.*

By contrast, for Morris Mini Clubman and onwards, things take a dramatic change with starter motor M35AK. However, as before, there are a variety of differences.

Firstly, there is Part No. 62925054 with the following features:

Model	62925054 M35AK
Origin	Australia
Pinion	54252212
Bracket (drive end)	54254175 or 54257544 or 62259275 (alloy)
Bush (drive end)	253316
Field Coils	62250983
Armature	62250970
Brushes	62251191
Bracket (commutator end)	62250973
Rear Bush	62250973
Spring retainer	Clip

Table 10.4.5 *M35AK starter motor.*

The AK starter motor is distinguished externally by the rear bracket which is now pressed steel. Inside, there is a face type commutator with spring loaded brushes held in a plastic housing.

Fig. 10.4.3 *Starter motor commutator and end bracket M35AK.*

Also specified for Morris Mini Clubman and Moke is 62925077 which is similar to the above except for drive end bracket 52457554 alone, and a different rear bush 62250486.

For Enfield manufacture, when power units reverted to UK CBU, the starter motor changed to M35J 25147 with UK parts:

Model	**251247 M35J**
Origin	UK
Pinion	54252212
Bracket (drive end)	54241172 (alloy)
Bush (drive end)	Not specified separately
Field Coils	54244772
Armature	54252038
Brushes	60600056
Bracket (commutator end)	54243579
Rear Bush	Not specified separately
Spring retainer	Clip

Table 10.4.6 *Part No. 25147.*

The through bolts enter from the front.

10.5 Wipers

Over 25 years in production, the wiper motor, arm and blade fitment for the Mini range has undergone some revisions and these lead to confusion for restorers looking for originality. Even Service Parts Lists show a confusing array of possibilities with mixtures of BMC and Lucas part numbers.

From about April 1968, wound field wiper motors were replaced with permanent magnet wiper motors, and wiper arms and blades changed from chrome to low-reflective finish (Morris Mini 2317; Mini Van 6052; Mini Deluxe 32801; Mini. Matic 1812; Cooper 'S' 4028 – SLS143). In late 1968 and early 1969 most models from BMC moved to MKII specification and higher end models were fitted with a two-speed wiper motor. At this time, electrical systems moved from positive earth to negative earth across the board for BMC Australian vehicles.

MODEL	LUCAS SPEC Part No	MOTOR TYPE	DWG	Angle	LUCAS SPEC ARM	PARTS BOOK	DWG	ARM LENGTH	ANGLE
ADO15	54071151 75385D	FIELD COILS	24G2344	120	54712030	54712030	6271047 24G1782	10.75	10 RH
ADO15						54711561			
YDO4,5,6		FIELD COILS	AYA9041	120		54711561	AYA9131	10.75	10 RH
YDO4,5,6		FIELD COILS			62710743	62710743	6271047	10.75	10 RH
YDO4,5,6		MAGNET			62710731		6271047	10.95	15 LH
YDO4,5,6	62710490	MAGNET	AYA9136	110	62710764	62710882	AYA9159 6271048	10.95	17 LH
2/YDO4,5,6	62071046	MAGNET	AYA9188	110					
YDO21	62971056	MAGNET	AYG9338	105		AYA9159	AYA9159	10.95	17 LH
YDO22,23	62071050	MAGNET	AYG9306	105		AYA9159	AYA9159	10.95	17 LH

MODEL	LUCAS SPEC BLADE	PARTS BOOK	DWG	BLADE LENGTH	FINISH	NOTES
ADO15	62710507	54711613	13H455 6271047	10 5/32	CHROME	Up to 3/68 Bayonet
ADO15		54711614				Rainbow
YDO4,5,6	62710507	62710507	AYA9143	10 5/32	LOW REFL	
YDO4,5,6	62710753		6271047	10 5/32	LOW REFL	3/67-10/67 FOR YDO5
YDO4,5,6	62710506		6271047	11 5/32	CHROME	
YDO4,5,6	62710744	62710744	AYA9158	11 5/32	LOW REFL	from 3/68
2/YDO4,5,6						NEG EARTH 3/69
YDO21	62710506	GWA9204	62710506	11 5/32	LOW REFL	SINGLE SPEED
YDO22,23	62710506	UPA34006 (10") GWB204 (11")	62710506	11 5/32	LOW REFL	TWO SPEED

Table 10.5.1. *Part numbers and specifications for Wiper Motors, arms and blades.*

10.6 Horn

10.6.1 Introduction

Sound waves in air are an example of longitudinal waves which are alternate compressions and expansions Δp between molecules at their equilibrium positions in the medium. The pressure changes involved in sound waves in air corresponds to about $\Delta p \approx 1$ Pa.

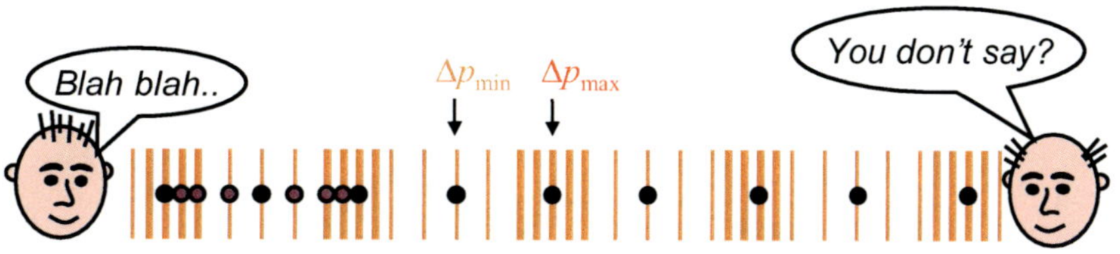

Fig. 10.6.1 *Sound waves represented as longitudinal compression waves in air.*

The velocity of sound in air is found from:

$$v = \sqrt{\frac{\gamma p}{\rho}} = \sqrt{\frac{\gamma RT}{M}}$$

In this equation, γ is the adiabatic index which is related to the ratios of the specific heats at constant pressure and constant volume. γ appeared earlier in Section 2.2 of Chapter 2. ρ is the density, p is the absolute pressure of the air and M is the molar mass (kg/mol). Pressure and density are related by the general gas equation $PV = nRT$ and so pressure and volume cancel, and the velocity of sound in air depends mainly on the absolute temperature T. The velocity of sound in air is about 340 m/s at 20°C.

The familiar Doppler effect that is experienced when a vehicle passes by with a horn or siren sounding arises from the increase in frequency as the wave fronts bunch up at the listener's ear upon approach, and their subsequent spreading out upon the vehicle's passing. Since frequency f and wavelength λ are related to wave velocity v,

$$v = f\lambda$$

then a decrease in wavelength results in an increase in frequency (approach); and an increase in wavelength thus is accompanied by a decrease in frequency (departure) since in both cases, the velocity of the wave remains constant.

The rate of energy transfer, or power, P per square metre of wave front carried by a sound wave is represented by its intensity I in W m^{-2} given by:

$$I = 2\pi^2 A^2 f^2 \rho v$$

In this equation, A is the lengthwise amplitude of the displacements of the molecules within the wave, f is the frequency in Hz and ρ is the air density and v the velocity of the wave.

The intensity *I* expressed as W m^{-2} is a linear measure. However, the ear can detect sounds which vary in intensity over a wide range and so for numerical convenience, a logarithmic scale is usually used where intensity is expressed in decibels and given the symbol β.

$$\beta = 10log_{10}\frac{I}{I_0}$$

I_o is the intensity at the threshold of hearing at 1 kHz, taken to be 1×10^{-12} W m^{-2}.

Some examples are:

Sound	Typical level in db
Whisper	20
Ordinary conversation	65
Train	90

Table 10.1 *Examples of sound intensity in decibels.*

When pressure differences are set up in a pipe, such as an organ pipe, standing waves are formed at the natural, or resonant frequencies of the pipe. The wavelength of the standing waves depends upon the length *L* of the pipe according to:

$$\lambda = \frac{4L}{n}$$

$$L = 1\lambda_1/4$$
$$L = 3\lambda_2/4$$
$$L = 5\lambda_3/4$$

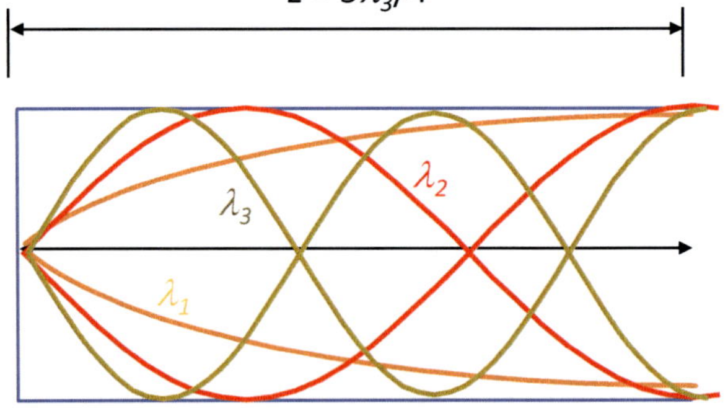

Fig. 10.6.2 *Standing waves in a pipe.*

Fig. 10.6.2 is a representation of the longitudinal or lengthwise displacements of molecules within the air along the length of the pipe for resultant wave.

Since frequency depends on wavelength, the frequency *f* is thus:

$$f = n\frac{v}{4L}$$

In this equation, *n* is an integer which takes values 1, 3, 5…. And each frequency generated is a harmonic, or overtone, of the fundamental frequency where at *n* = 1. As the length increases, the frequency decreases. Thus, a low note horn has a longer "trumpet" than a high note horn.

For the Lucas Windtone low note horn, the frequency is specified at 390-405 Hz and the output is 114-118 db. For a high note horn, the frequency is specified at 488-503 Hz, and the output is 114-118 db.

For a high note horn, measurements show that the trumpet length is ≈180 mm. With v = 340 m/s, then we might expect the horn to produce overtones as shown in Table 10.2. Each overtone will have its own amplitude and the shape of the horn, in particular, the exit flare, determines which overtones dominate. A mellow sound occurs when only one or two lower overtones are heard. A sharp sound occurs when higher frequency overtones dominate.

n	CalculatedHz	Measured
1	472	450
2	944	950
3	1417	1400
4	1889	1880
5	2361	2450
6	2833	2800
7	3306	3300

Table 10.6.2 *Calculated and measured harmonics for Lucas Windtone high note horn.*

The relative amplitudes and actual frequencies in a sound wave can be measured with a microphone and an oscilloscope which has a frequency analysis function. The sound waves are converted to electrical signals by the microphone, and the oscilloscope extracts out each individual harmonic frequency and determines its amplitude using a technique called Fourier analysis. The output is a plot of amplitude vs frequency. A relatively pure sound is heard when there are only a few component harmonics with one or two at larger amplitudes. Measurements on an actual horn are shown in Fig. 10.6.3 and tabulated in Table 10.6.2.

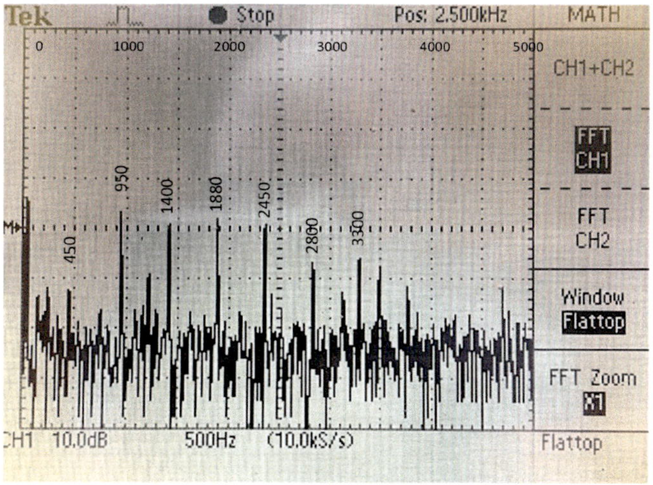

Fig. 10.6.3 *Fourier analysis of the sound from a Lucas Windtone high note horn.*

It would seem here that the higher overtones are more dominant than the fundamental.

10.6.2 The Lucas Windtone Horn

In the Lucas Windtone horn, air vibrates in the trumpet as a result of the vibration of a metal diaphragm driven by an electromagnet which alternately is powered on and off mechanically. The length of the trumpet determines the pitch of the sound. There is a marking on the horn just inside the opening for "L" and "H" frequency pitches. For vehicles with two horns, Lucas service documentation states:

> *"When used in blended pairs, one high note and one low, and operated simultaneously, WT614's give a pleasantly mellow and harmonious signal. You will notice that the horn trumpet is coiled so as to obtain the desired length in the minimum of space. The high note model has the shorter coiled tube and the letter ' H ' is imprinted inside the trumpet flare. The letter' L ' is used to distinguish the low note version. As a point of interest to the musicians among you, the notes differ in pitch by the interval of a major third."*

As can be seen in Fig. 10.6.4, there is a metal diaphragm which vibrates according to the motion of an electromagnet, and these vibrations cause resonant sound waves in the air horn. The steel diaphragm often rusts around the edges where water can seep in past the gaskets. However, the interior of the horn is usually very well protected from the elements.

Fig. 10.6.4 *Inside the horn showing spring-loaded contacts. The metal diaphragm, rust has formed from moisture seeping in from the edge.*

Battery voltage is attached to one terminal and the horn push completes the circuit via the vehicle earth. An electromagnet inside the horn is activated and deflects the diaphragm, but in doing so, the circuit to the electromagnet is broken and the spring pushes the diaphragm back whereupon the cycle repeats. Over time, the contacts can wear which affects the mechanical action of the electromagnet, and so there is an adjustment screw available to compensate. The adjustment screw is not the large screw and locknut in the middle of the horn, but a flat slotted screw over to the side. The larger slotted core and locknut is used to adjust the air gap inside the horn and should not require adjustment.

The two halves of the horn are not designed to be taken apart, they being riveted together. If it is necessary to dissemble the horn, then some care needs to be taken not to disturb the mechanical adjustments.

If the horn makes no sound at all, and there appears no continuity between the terminals, then do not assume the problem is internal. Seemingly intact connections at the exterior Lucar connectors may become open circuit (that is, from the connector to the securing rivet) via electrolytic corrosion. It is therefore important to check continuity both at the rivets which secure the Lucar blades, and then at the connector blades themselves. The DC resistance of a 12V horn should be about 0.5Ω when measured with an ohmmeter.

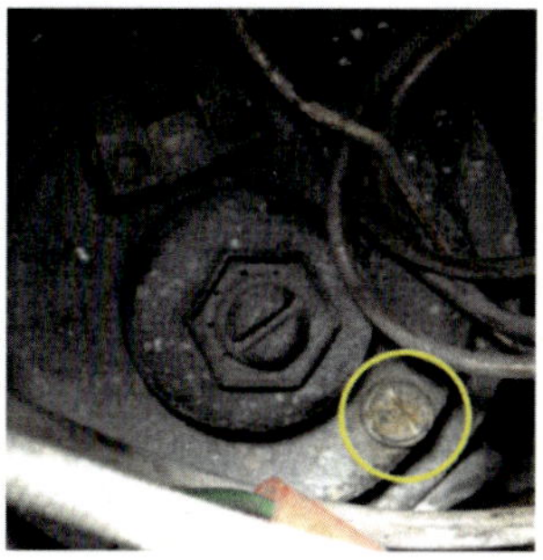

Fig. 10.6.5 *Adjustment screw (lower right). Adjust for about 4A when horn is sounding.*

10.6.2 Minor Adjustment

Adjustment of the horn must be made using an ammeter. For a typical Windtone horn, Lucas recommends the current drawn by a single horn should be about 6 amps. However, all this current is coming through the horn push contact and not through a relay. In practice, a little under 4 amps usually produces a satisfactory output on a single horn. With dual horns, the total current through the horn push contact should be kept at about 6 amps, probably about as much as the horn push contacts can take without too much trouble. The purple wire is live, and the purple and black wire goes to the horn push contact and thence to earth.

10.6.3 Major Adjustment

The relevant section of the Lucas service literature is reproduced below.

a) Slacken centre core locking nut and lightly rotate the centre core clockwise until it just touches the armature – when opposition to further rotation will be felt.
b) Rotate the centre core anticlockwise for one and a half turns and lightly tighten the locking nut.
c) Connect the horn in series with an ammeter and horn push and apply 12V to the circuit.
d) Depress the horn push and note the operating current indicated by the ammeter. If necessary, turn the contact breaker adjustment screw until the correct current is obtained – turning the screw clockwise to increase the current, or anticlockwise to decrease it. Release the horn push.
e) Set the circuit voltage to the upper limit, 14V for 12 V units.
f) Depress the horn push and turn the centre core clockwise until the armature just impacts against the core – this condition being indicated by the harsh note emitted.
g) Turn the centre core anti-clockwise (1/8 to ¼ of a turn) until a good clear note of adequate volume is emitted. Release the horn push.
h) Tighten the centre core locking nut.
i) Depress the horn push and observe the ammeter and, if necessary, turn the contact breaker adjustment screw until the correct current is obtained.

10.6.4 Applications

Horns should be mounted on their original bracket. The flex in the bracket makes a substantial difference to the volume and quality of sound. Service Parts Lists shows the following horn fitments:

Model	Part No.	Tone	
ADO15, ADO15V, YDO4, YDO4/V	70141, 70159 62068001B 62968006 62068055	High Note	Lucas
YDO4, YDO4/V	AYB9083	High Note	RVB

YDO5, ADO50, YDO6, ADO16	62068003D	High Note	Lucas
ADO50, YDO6, ADO16	62068004B 62068006	Low Note	Lucas
ADO17	62968008 62068009 62068031	Low Note	Lucas
ADO17	62968006	High Note	Lucas
YDO7, YDO18	AYB9083	High Note	RVB
YDO21, YDO22	62068056	Low Note	Lucas
YDO21, YDO22 YDO21, YDO22 (SS LS)	AYH9712 AYB9083	Low Note High Note	RVB
YDO9, YDO15	62068008 62068009	High Note Low Note	Lucas
YDO19 Tasman	62068040		Lucas
YDO19 Kimberley	62068008 62068009	High Note Low Note	Lucas
YDO24	AYH9726	Low Note	RVB
YDO26, YDO27, YDO28	AYD9070 AYD9069	Low Note High Note	RVB

Table 10.6.3 *Horn application*

Note that for some models, an RVB[28] horn is fitted, not Lucas. RVB horns are adjusted in a similar manner to that described above. In practice, different models of both Lucas and RVB horns have different external appearances, and some examples are given below.

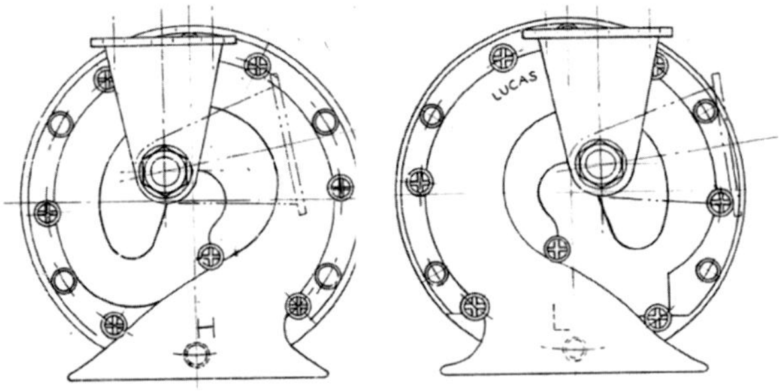

Fig. 10.6.4 *Lucas Windtone horn, high note, 62068006, low note, 62068008 (MYH4198).*

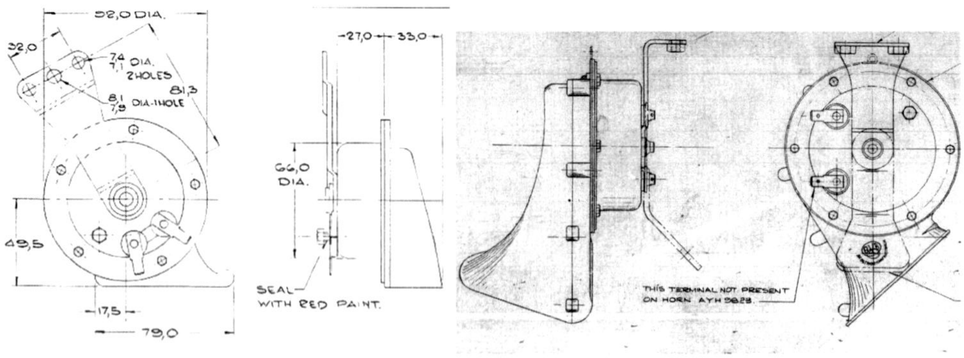

Fig. 10.6.5 *RVB Horn, High note, AYB9083. Low note, AYH9712.*

[28] named after the company's founder Roy V. Butler

10.7 The Astor Diamond Dot Car Radio

10.7.1 Introduction

The Astor Diamond Dot car radio was offered by several manufacturers during the 1960s and 70s. At General Motors Holden, it was marketed as the Air Chief. At BMC, it was simply the BMC All Transistor.

In general, there were the 8, 11 and 16 transistor versions. The 11 transistor was probably the most popular and featured several innovative items, the most well-known being the crackle cutter switch set off by the blue plastic "diamond dot" in the facia panel.

> "THIS BMC PUSH BUTTON RADIO NOW FEATURES A "CRACKLE CUTTER" SWITCH which is located behind the manual tuning knob. This effectively cuts out high frequency interference and allows the listener to continue enjoying the selected programme. For instant selection of stations they have 5 push buttons which can be set to any 5 preselected stations."

Fig. 10.7.1 *Astor Diamond Dot car radio illuminated dial with crackle cutter enabled.*

The 11 transistor model has 8W audio output, while the lesser 8 transistor 3W. The 11 transistor radio feature an audio power amplifier of the push-pull design which gives lower power consumption and higher output compared to single power transistor models. The semiconductors were made in Australia with the brand name Anodeon.

These car radios are now considered classic items and sell at collector prices. Despite the rarity of the components, they can be easily repaired with some understanding of how the circuit works.

At first sight, the circuit diagram for an Astor Diamond Dot radio appears overwhelming, but when broken down into component functions, becomes much less so.

Fig. 10.7.2 shows the main sections of the circuit. In the radio frequency (RF) part of the circuit, radio waves are received by the aerial and a particular station is tuned in using the manual or push button tuner. There is an internal oscillator OSC, whose output is mixed with the RF signal to production an intermediate frequency (IF) signal, which is then passed to an audio amplifier for output to the speaker. The power supply to the circuit is taken from the vehicle battery and is filtered to remove interference from the vehicle's electrical system.

10.7.2 Radio Frequency Receiver

The aerial receives a multitude of radio waves of different frequencies. Variable inductance coils are adjusted by a front panel control by the listener to match the resonant frequency of the aerial receiver circuit to with that of the desired station. That is, the aerial receiver circuit is thus tuned to resonate with the desired carrier wave radio frequency (RF). For AM radio, in the medium wave band (MW), the spectrum of carrier frequencies used is in the range 500 kHz to about 1600 kHz.

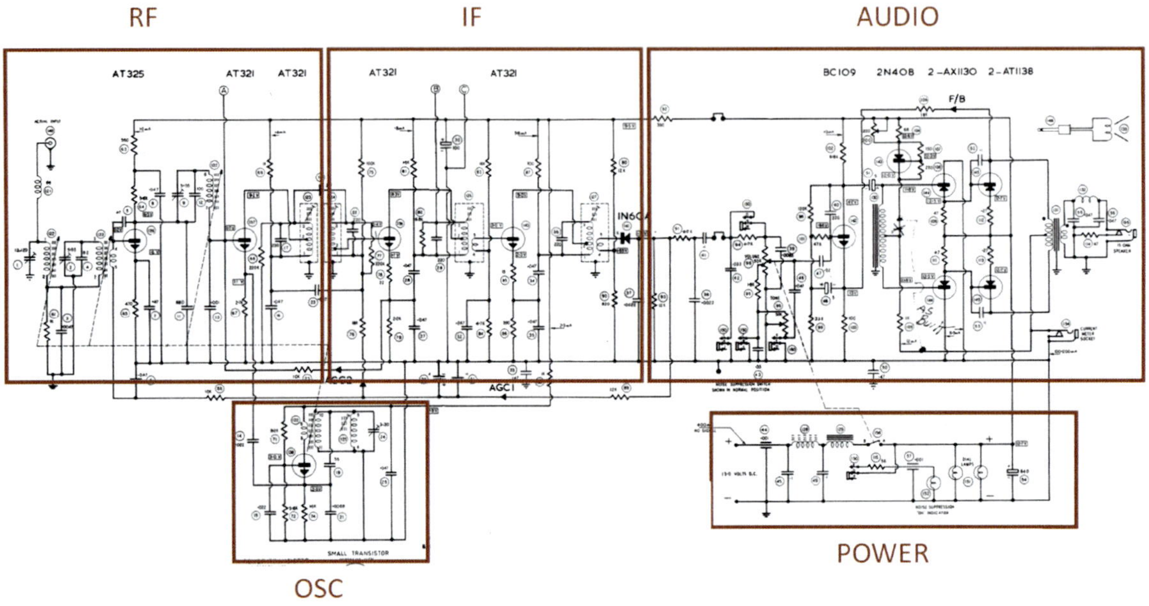

Fig. 10.7.2 *Circuit diagram main block functions.*[29]

While the frequency of the carrier wave for a particular station is constant, the amplitude of the wave is not. The amplitude varies according to the audio signal being transmitted. If a single tone is being transmitted, then the amplitude of the carrier wave oscillates in a smooth sinusoidal manner. If more complex sounds are being transmitted, then the amplitude of the carrier wave varies in a more complicated manner, but mathematics shows that whatever the form of the amplitude variation, it can be expressed as the sum of a series of sine waves (harmonics) of difference frequencies. The variation in amplitude of the carrier wave is called modulation. That is, the amplitude of the RF carrier wave is modulated, and this modulation, at audio frequency, carries the audio signal being transmitted.

The RF amplifier has a small amount of gain associated with it, but its main purpose is to match the high impedance of the aerial with the low input impedance of the mixer transistor.

[29] Acknowledgement Kevin Chant

10.7.3 Oscillator and Mixer

AM radios generally employ the super heterodyne principle in which the modulated RF, once received by the receiver circuit, is converted to a modulated intermediate frequency, IF, amplified and then presented to an audio power amplifier for output to the speaker. The process by which the received modulated carrier is converted to a single IF is the basis for the term "heterodyne".

An internal oscillator is set to oscillate at a fixed frequency above the tuned-in RF. The amplitude of the oscillator output is constant. The difference in frequency between the oscillator frequency and the RF is the intermediate frequency, IF. The IF is typically 455 kHz. The oscillator tracks the RF tuning circuit mechanically via the front panel tuning control so that it is always IF (i.e. 455 kHz) above the tuned-in RF.

The output of the RF amplifier is fed into the base of the mixer transistor. The output of the oscillator is fed into the emitter of the mixer transistor. The combination of the two frequencies (one amplitude modulated RF, and the fixed amplitude oscillator frequency), alters the voltage between the base and the emitter, which, via transistor action, results in the voltage of the collector of the mixer transistor oscillating at the intermediate frequency IF but also carrying the amplitude modulation of the original RF.

The process is the same as the beat frequency that can be heard when tuning musical instruments when one hears two sounds waves that are very close in frequency. The frequency of the beat is the difference in the two sound wave frequencies. In this case, the beat frequency, (IF) is fixed, but because the amplitude of the RF varies, the *amplitude* of the beat frequency, or IF, also varies. In this way, the amplitude-modulated RF is converted into amplitude-modulated IF.

For example, if the amplitude-modulated RF of the tuned-in station is 1000 kHz, the oscillator output is then 1455 kHz. If the RF is 1200 kHz, the oscillator output is 1655 kHz. The output from the mixer is always a modulated 455 kHz signal.

10.7.4 IF Amplifier

The IF amplifier is a tuned single-frequency amplifier. The intention is to provide a high quality amplifier which locks into a specific frequency (IF) and rejects all other frequencies. An alternative approach would be to just amplify the tuned-in RF carrier wave no matter what its frequency, but it is easier and cheaper to build a low-noise *single-frequency* amplifier rather than a broadband amplifier. The intermediate frequency is much greater than the bandwidth of the audio signal and so its amplitude modulation is well able to carry faithfully the original audio signal. The IF amplifier itself is a fixed frequency tuned circuit with adjustable elements to optimise its response at the IF. The tuning stages of the IF amplifier serve to select out the modulated IF signal and reject signals of other frequencies which would be considered noise. This design is sometimes termed a "lock-in" amplifier. However, if noise finds its way into the RF circuit, it will be mixed in with the oscillator and carried along with the audio and amplified.

A diode, referred to as the "detector", passes the top half of the IF signal through to the audio amplifier. This rectified and amplitude modulated signal is immediately sent to a low pass filter which blocks the IF and passes the remainder, the audio signal, through to the audio amplifier. At this point, it should be realised that the audio amplifier receives a signal of mixed frequencies and amplitude. The 455kHz IF frequency has been stripped off.

The output from the detector diode is heavily filtered (low pass) and approaches a DC level which depends upon the strength of the audio signal. This is fed back into the base of the RF amplifier transistor, and also the mixer, which decreases the bias voltage and decreases the gain. When the radio is receiving a weak signal, the RF and mixer transistors gains are maxed out. When the radio is receiving a strong signal, the gain is reduced. This automatic gain control (AGC) minimises variation in sound level (i.e. volume) when reception fades, or between close and distant radio stations, so that the listener is not always adjusting the volume control for comfortable hearing.

10.7.5 Audio Amplifier

At just past the detector diode, the signal is an audio signal comprising a mix of audio frequencies and amplitudes representing the sound being broadcast. This signal might range from a single tone to a complicated waveform representing the harmonic components of speech or music. The audio signal typically contains frequency components in the range say 20 Hz to about 15 kHz. It is the superposition of these amplitudes of the component frequencies that makes up the total audio signal being amplified and then sent to the speaker. What has happened here is that the amplitude modulation of the IF (455 kHz) has been converted to multi-frequency audio signal whose overall sum is that of the amplitudes of its component harmonics (in the range 20 Hz to 15 kHz).

The audio amplifier can be thought of as containing a voltage amplifier (gain adjusted by the front panel volume control) and a power amplifier (as distinct from a *voltage* amplifier). The power amplifier has little voltage gain but is able to provide enough current to drive the speaker. These amplifiers operate in the audio frequency range.

For the power amplifier, the audio signal is passed through the primary of an input transformer which has the effect of producing two signals, one the inverse of the other. The output transformer later combines the two signals for presentation to the speaker.

Transformer coupling for both input to, and output from, the power amplifier is required because both transistors are PNP. The input signal is split into two copies by driver transformer (130), one out of phase with the other, so that only one power transistor is on at any one time. This enables transistors of the same type to be used in push-pull capacities. By contrast, in more modern complementary push-pull configurations, PNP and NPN transistors are used which perform the same function – each transistor to amplify half the signal. This is termed a Class B amplifier. A Class A amplifier uses a biased single transistor as the power amplifier and this type of circuit can be found in the Astor 8 transistor range.

Making a circuit this way means that the power consumed by the radio is less when the volume is low, and more when the volume is greater. Amplifiers with a single transistor (Class A) operated at their bias currents continuously and so consume power even in the absence of any input signal. Class B amplifiers (push-pull) only amplify a signal when it is present. Class AB signifies that the circuit is mainly Class B, but a small amount of transistor bias is applied (small collector current) in the absence of any signal to eliminate cross-over distortion as one transistor turns on and the other turns off.

The speaker transformer (131) then combines the two half output waveforms for presentation to the speaker coil via transformer secondary. The speaker transformer also acts as an impedance match between the speaker coil and the power transistors.

The power amplifier circuit has transistor bias applied via adjustable pot (107) to eliminate cross over distortion that arises due to the small "turn-on" voltage at the base of each transistor. would have arisen if the circuit was class B only. The circuit is thus termed Class AB, incorporating Class A action at small signal levels and Class B dominating at larger output levels. For the Astor 11 transistor, the collector bias current is specified as 150mA with V_{cc}=13V or 120 mA at V_{cc} = 12V.

There is negative feedback to the driver transistors via resistor (106) to take into account any positive feedback that might arise from ripple on the transformer primary circuit.

This is a power amplifier, not a voltage amplifier. The intention is to transmit maximum power to the speaker. As a consequence, the power transistors are relatively high current devices and are mounted on a heat sink.

10.7.6 Speaker Transformer

The output specification for the Astor 11 transistor radio is 8W. Assuming 80% transformer efficiency, the average input power to the transformer is

$$P_{in} = \frac{8}{0.8} = 10W$$

Allowing a 25% factor of safety, we can let P_{in} be:

$$P_{in} = 10 + 2.5 = 12.5W$$

This is the required input power to the speaker transformer. Since each transistor supplies one half of the output waveform, each transistor must then be able to deliver 6.25W[30].

For any AC signal comprising sinusoidal voltage and current, the average power transmitted by the signal is found from the rms values of these quantities:

$$P_{av} = \frac{V_o}{\sqrt{2}}\frac{I_o}{\sqrt{2}} = 0.5V_oI_o$$

---

[30] The average power is also called the "rms" power. It is only called "rms" power because it is calculated from the rms values of the voltage and current. It represents the power which must be supplied to the circuit and is usually expressed in units of VA. The "active" power, the power actually consumed, is the rms power multiplied by a power factor that takes into account power alternately supplied to and received from the circuit arising from reactive components. The active power is usually expressed in W.

Note, to determine the efficiency of the transformer, we note that the voltage Vcc is constant, and not sinusoidal. The current is sinusoidal. Hence, the average power drawn from the power supply must be

$$P_{av} = V_o\frac{I_o}{\sqrt{2}}$$

Where V_o and I_o are the peak values of the sinusoidal voltage and current, in this case, +12V and -12V.

On the primary side of the transformer, if the peak value of voltage V_o is V_{cc} = 12V, then the peak value of the current is:

$$P_{av} = 0.5V_o I_o$$

$$12.5 = 0.5(12)I_o$$

$$I_o = 2.1A$$

Even though there are two transistors, they each are required to have this current carrying capacity. Each transistor conducts on each half wave form, the significance of which is that each transistor supplies 6.25W of power – the same peak primary current appearing for only *half the time* on each compared to the case where a single transistor is handling a complete waveform.

We can treat the speaker as being a load impedance of say 15Ω. The power output is 8W, but as we calculated above, we can make some allowances for inefficiencies in the components involved leading to a required input power to the transformer of 12.5W.

In the *primary* circuit, if the speaker transformer were to be replaced with an equivalent load impedance Z_L, then the required value of Z_L is:

$$Z_L = \frac{12}{2.1} \approx 6\Omega$$

The speaker transformer for the Class AB amplifier in the 11 transistor Astor car radio has 200 turns of 0.5mm dia copper write on the speaker output side, and 130 double wound turns of the same sized wire on the input side. That is, each input winding consists of 130 turns, the windings having a centre tap which connects to ground. For each winding, the turns ratio N_s/N_p = 1.7.

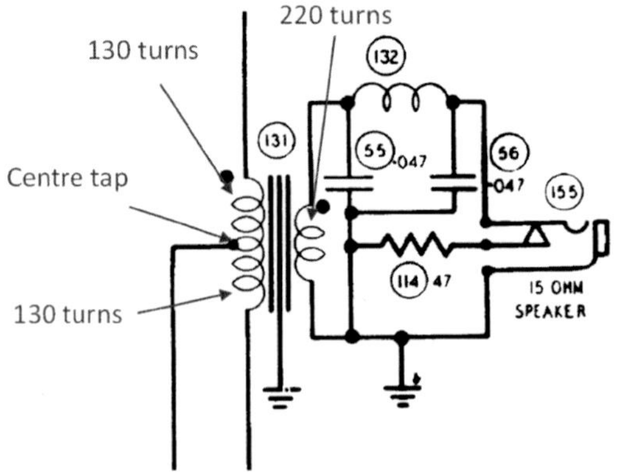

Fig. 10.7.3 *Speaker transformer circuit.*

If n is the turns ratio N_s/N_p, then the impedance ratio is the square of this:

$$n^2 = \frac{Z_s}{Z_p}$$

For maximum power transfer between the transformer secondary and speaker, the impedance of the secondary must equal that of the speaker, and so for the 15Ω impedance speaker, the primary impedance must therefore be:

$$\begin{aligned} Z_p &= \frac{15}{1.7^2} \\ &= 5.2\Omega \end{aligned}$$

This is fairly close to the estimate 6Ω made above. That is, the 15Ω speaker results in a 6Ω reflected impedance for the load on the power resistors. DC measurements on an actual speaker transformer show that the primary has a total *resistance* (each winding) of 1.1Ω and a secondary resistance of 1.6Ω. DC measurements on a nominal 15Ω speaker indicates a resistance of 11.2Ω.

10.7.7 Fault Finding

In common with many older electronic devices, the most common component to fail are the electrolytic capacitors. The dielectric between the plates of these capacitors evaporates over time. It is an easy matter to replace them with modern equivalents.

While there are only a small number of electrolytic capacitors to replace, there are a multitude of other types which should be treated with suspicion and care. Paper capacitors can also fail over time.

Ceramic and polystyrene capacitors are usually reliable, but one might notice that the polystyrene capacitors are held off the circuit board with tubular stand-offs. This is because they are susceptible to irreversible heat damage from soldering and care should be taken when replacing any nearby components. Polystyrene capacitors can be obtained new, but supply is limited and at relatively high cost. They can be replaced by more modern polyester types.

Nearly all of the resistors are of the 0.5W carbon type. While it is tempting to replace the resistors with modern equivalents, they should only be done so if any are suspect and then replaced with carbon types, especially in the RF, OSC and IF circuits. Modern metal film resistors may have undesirable inductances.

Apart from component failure, the most common cause of overall failure are burnt out power transistors arising from connecting a speaker of too low impedance, or shorting the speaker wires together. The resulting heavy current through the power transistors will almost certainly burn them out, and possibly burn out the output transformer as well. These power transistors are PNP Germanium types and not so easily found as spare parts.

Fault finding and repair occurs in reverse order to signal input in that attention is given to the following sections in order of: Audio Amplifier, IF Amplifier, Oscillator, RF Amplifier.

Item	Notes
Attenuator	Details of this are not so easy to find in the Astor service literature. One document refers to this as a 4nF ceramic capacitor in series with a 10K resistor.
Dummy aerial	The Astor literature refers to this as a 65pF ceramic capacitor in serial with the signal generator output and experience shows that this works well, but can give an overly strong signal in the last stages of circuit alignment.
Coupling capacitor	0.01uF polyester capacitor for introducing IF alignment signal

Plastic core adjustment rod	Transformer core adjustment cannot be reliably set with a metallic tip. A flat bladed plastic tip is preferred.
Signal generator	The signal generator must have the capability of an amplitude modulated sinusoidal output. Typical modulation required is 30% of the generator output. Frequency range in the 100 kHz – 1mHz region. A frequency sweep function is not essential. The alignment procedure requires frequency adjustment of the signal generator frequency output and any instrument that provides a push-button customizable output is desired.
Oscilloscope	A two channel oscilloscope is required – one for measuring audio output, the other, for RF and IF fault finding. Since the audio frequencies are much lower than the RF and IF frequencies, two separate oscilloscopes are desirable so as to avoid having to change the time base.
Output meter or speaker	A 15Ω speaker is required for this radio to operate. Some means of measuring the audio output is required. This can be simply the volume of sound, or, an oscilloscope display of the speaker signal – with the speaker connected. It is important to have the output loaded with the speaker impedance when measurements are made.

Table 10.7.1 *Equipment needed for radio repair and adjustment.*

Many of the adjustments are done for maximum output at the speaker. This can be as simple as just listening for maximum volume, or, preferably, connecting an oscilloscope to the speaker connection, with the speaker attached, and observing the amplitude of the output waveforms. This has the advantage of displaying any signal distortion although high frequency harmonics are usually audible.

10.7.7.1 Audio Amplifier

The operation of the audio amplifier can be tested by feeding in a 400 Hz sinusoidal signal at the test link following the coupling capacitor (41). There are two test links on the circuit board. Looking from the component side, the audio test link is the one positioned towards the centre of the circuit board. The other test link (towards the edge of the board) is a power feed to the IF and RF circuits.

Most problems in this area are found to be failure of the driver or power transistors. If any of these fail, then a short circuit condition can be obtained and it is important that when fault finding, a current limiting power supply set at about 1.0 A should be used to avoid causing component failure. When operating at low volume, and with the dial indicator lamp connected, the current consumption of the radio unit should be no more than 400 mA.

Both test links can be removed so that a low level audio signal, of say 400Hz, can applied to the pin closest to the RF tuner mechanism – the audio amplifier input. With the volume control at mid-way, this signal should be heard at the speaker. If not, then the signal should appear at the electrolytic capacitor 51. If so, and there is still no audio output, the base and emitter of the driver transistors 144 should be checked. Lastly the base and collector voltages of the power transistors 145 are checked. It is sometimes found that the driver transistors are the problem, even to the extent of the terminal leads being broken from fatigue inside the plastic stand-off tubes. Driver and power transistors are to be replaced as a matched h_{fe} pair. If this is not done, then the output may be distorted at higher volumes as one transistor saturates before the other.

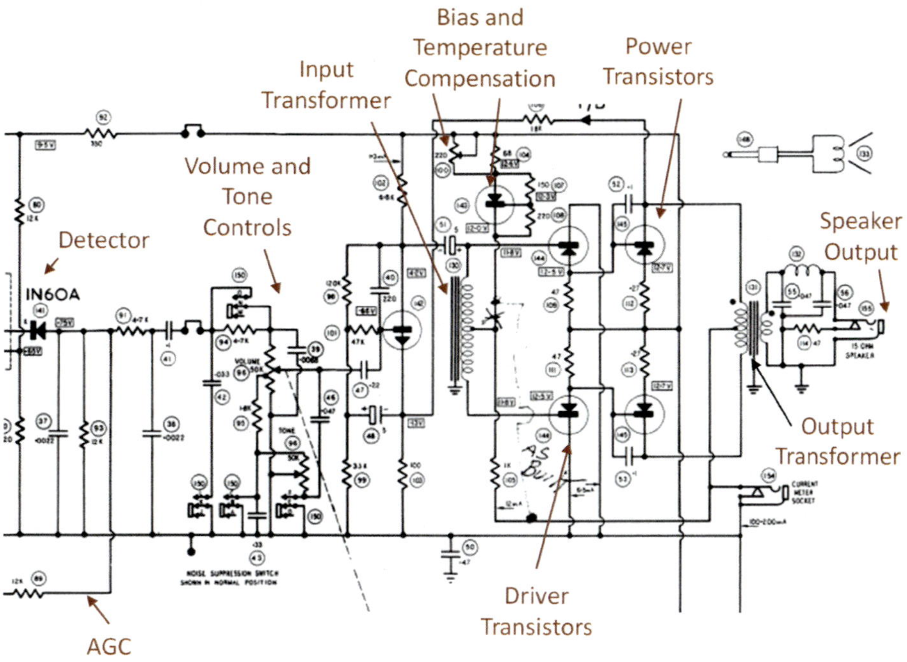

Fig. 10.7.4 *Audio Amplifier.*

The bias current for the power transistors should be adjusted according to the service instructions and is usually within the range of 100 to 150 mA. If it is found that no adjustment is possible, and the current remains in excess of about 200 mA, then the trimmer potentiometer is likely to be the cause. Excessive current can also be caused by failure of one of the driver transistors. With a test wave form at the input test pin, the volume control should alter the amplitude of the signal at each power transistor base. If replacing transistor 142 with a BC109, then select for low value of h_{fe}. If the audio amplifier gain is excessive, then R103 can be increased to about 150Ω.

10.7.7.2 IF Signal Amplification

Once the audio amplifier operation has been established, it is necessary to give attention to the high frequency side of the circuit.

There are four IF transformers that require adjustment, or tuning in, to the IF frequency of 455kHz. The main tuning dial should be at the top end of the frequency range so as to avoid interference with the 455 kHz test signal. A fixed frequency 455 kHz signal modulated at 400 Hz at about 30% is fed into the mixer transistor base (test point A) via a 0.01 μF coupling capacitor. The 400 Hz audio output is monitored for greatest amplitude as each of the ferrite cores of each IF transformer is adjusted in sequence. The cores are adjusted from the one closest to the power amplifier back towards the mixer (i.e. 4, 3, 2, 1). Interaction between the adjustments will mean that the procedure will have to be repeated until there is no more to be gained at the output. At final adjustment, the cores should be a reasonable height in the former, not bottoming out or excessively high. The input signal should be reduced as adjustments are made so as the smallest possible input can still provide a detectable output wave on the oscilloscope. Great care should be taken to extract the maximum signal from the IF stage. Once adjusted, the IF amplifier can be left alone while broadcast tuning is done.

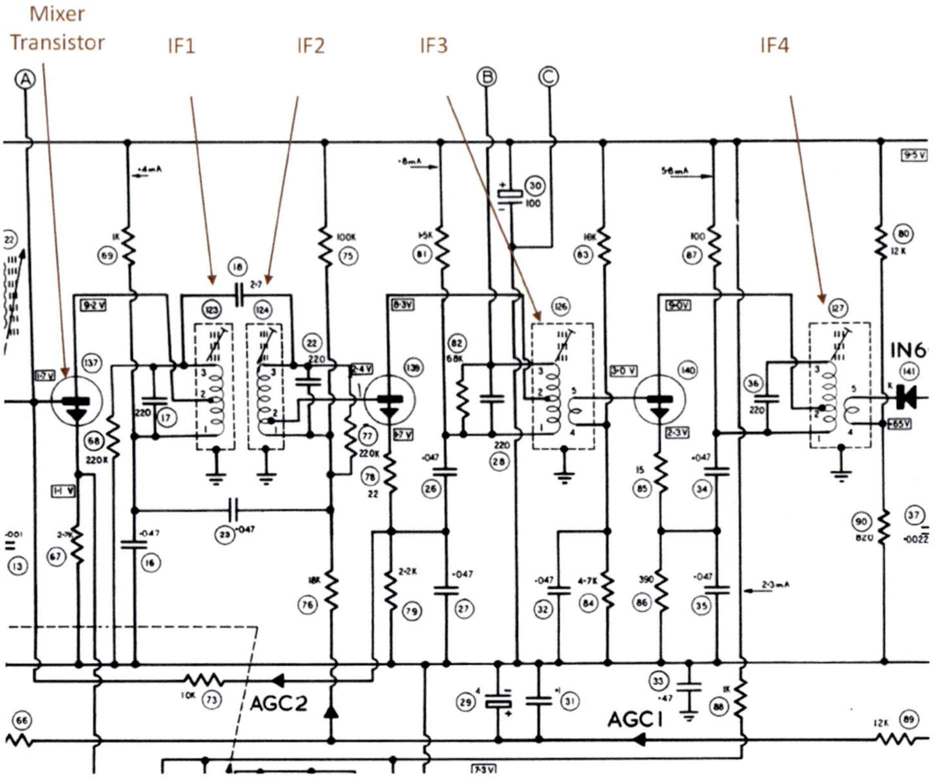

Fig. 10.7.5 *Mixer and IF amplifier.*

It is important that the signal generator not drift in frequency during the adjustment. Typically, the input signal amplitude is made as low as possible so that any output does not saturate, and that the effect of any automatic gain control is minimised.

Ferrite cores are delicate items and easily cracked. For cores which have not been adjusted for some decades, there is the possibility that they could be stuck in the former. In this case, some penetrating oil might need to be applied overnight to free them. If an attempt is made to free a stuck core with the tip of a screwdriver, there is the possibility of chipping off small amounts of the core which will then jam the thread and make matters worse. Cores should be adjusted with a non-metallic tool so as to protect the core, and to remove any influence on the inductance of the device that would occur if a metallic tip were used.

10.7.7.3 Oscillator

The oscillator output frequency tracks the RF carrier wave frequency mechanically via the tuning gang mechanism. It is arranged so that the oscillator output is at a frequency IF (455 kHz) above the RF carrier wave frequency. The IF amplifier is looking for an input of 455 kHz, this being the difference between the RF amplifier output and the oscillator output.

The oscillator is a common-base Colpitts design where a resonance is set up between the charging and discharging of two capacitors and an inductor. The resonant frequency in Hz is given by:

$$f_o = \frac{1}{2\pi\sqrt{LC}}$$

where C is the series combination of C19 and C21, modified by the tuning components C24 and L125.

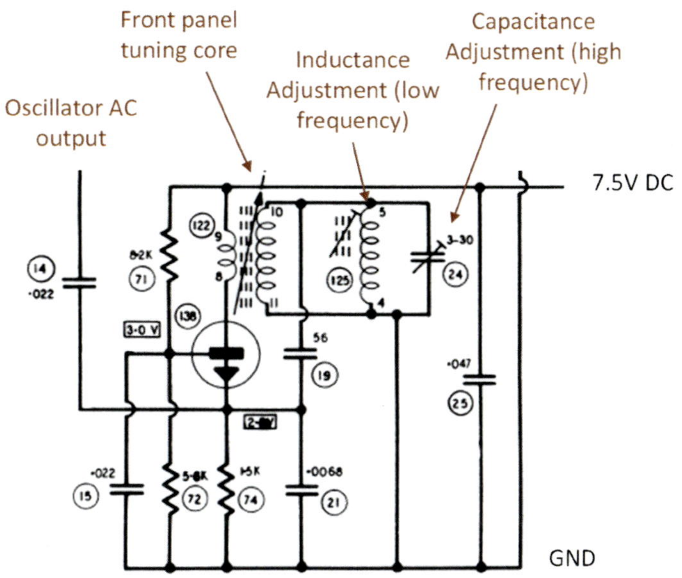

Fig. 10.7.6 *Oscillator.*

With no input signal to the aerial socket, and with the tuning cores set out of their coils (front panel tuning to top of the dial), the presence of an oscillator output should be checked with an oscilloscope. The oscillator output can be seen at the emitter of (138) or capacitor C14. If there is no oscillation, then the cause must be found, and the problem remedied. There are only a handful of components to consider, the first being the transistor 138. The DC voltages at each pin of the transistor should be checked against those on the circuit diagram and should be within about 0.5V of those indicated. Replacement of the associated capacitors and resistors should be considered if the transistor is tested and found to be working. There is no use in proceeding with any other adjustment until there is an oscillator signal.

The iron core of the inductor (L125) should be adjusted initially so that it is about level with the top of the tube. Then C24 is adjusted to give a 2080 kHz output (455 kHz plus the 1625 test carrier to be used in RF alignment).

Note, adjustment of C24 influences the high frequency end of the response at this point. Later, when a low frequency tuning is required, the ferrite core coil L125 is used for low frequency adjustment.

10.7.7.4 RF Amplifier

Tuning the RF and oscillator stages of the radio is the most difficult part of the service operation and requires a feeling for what is supposed to happen as various components are adjusted for best results.

In this procedure, a test signal is introduced into the aerial. The test signal is modulated at 400 Hz as before at about 30%. Connection of the test signal can be made in several ways. In series with a 68pF ceramic capacitor attached to the centre pin of the aerial socket; via an IEC specification dummy aerial, or, with an aerial connected, and a loop "transmitter" from the signal generator placed close by. The direct connection via a 68pF works well, but can give an overly strong signal at the last stages of adjustment. Using a loop transmitter has the advantage of loading the receiver circuit with the impedance of the actual aerial that will be used in service.

An attenuator is connected to pins B and C. The attenuator is a 4nF ceramic capacitor in series with a 10K resistor. The tone control is turned to maximum treble, and the volume control to maximum. The dial face, front panel indicator pointer and black backing panel are to be removed to allow access to the ferrite tuning core adjustment rods.

The front panel tuning control is set to the high end of the frequency range against the mechanical stop, and then backed off a little to reduce mechanical strain. The ferrite cores of the front panel tuning control are wound out so that about 3mm of the threaded adjustment rods are protruding from the front panel metalwork.

At a signal generator output of 1625 kHz, at moderate amplitude, the oscillator (previously set to 2080 kHz) is adjusted slightly for maximum output at the speaker. The amplitude of the modulated test signal may need to be reduced as optimum adjustment of the oscillator is achieved.

The aerial trimmer capacitors (1) (hand adjustment) and (2) (screwdriver adjustment) are adjusted for maximum output, and also the RF trimmer capacitor (9). Ideally, these trimmer capacitors should not be hard up against any full scale stop or wound out to a loose condition.

The front panel tuning control then needs to be positioned at about 1000 kHz position, but this is done mechanically by using a calibrated spacer. A spacer is not absolutely necessary. The control can be adjusted so that the gap at the tuning gang mechanism is 0.39" (10 mm) measured with a vernier calliper or cardboard section. The dial indicator mechanism should be a little to the right of centre of travel at this point. A mark should be made on the metal work to indicate the position since the front panel tuning will be altered in the next step but needs to be brought back to this reference position.

The signal generator is adjusted to 1000 Hz. It is unlikely that the front panel tuning will be tuned into this frequency since the ferrite cores are in their outer most position. Thus, it is easiest to adjust the front panel turning knob downwards in the frequency range until the test signal is picked up and audible at the speaker. The front panel tuning control can then be "walked" up to the reference position adjusting the OSC, RF and Aerial 1 and Aerial 2 cores in that order. The OSC and RF are the most influential adjustments at this point.

Once there is an audible signal when the dial indicator is at the reference 1000 kHz position, all four cores should be adjusted for maximum output. It will be necessary to reduce the amplitude of the test signal from the signal generator as tuning in proceeds.

This adjustment requires some skill and patience since adjusting one core affects the optimum position of the others. The signal generator should be at a minimum output at the completion of the adjustment. If the signal generator, or test signal, is too large, then the automatic gain control of the RF amplifier will smooth out the peaks in the response and make identification of the maximum output difficult.

The test signal frequency now adjusted to 600 kHz and the front panel tuner is then adjusted to the lower end of the frequency range to pick up the signal. The ferrite core of the oscillator inductor (125) is now adjusted for maximum output. Adjusting this core affects the response of the circuit and so it is necessary to walk or rock the font panel frequency control a little to find the new maximum.

If the oscillator inductor has been adjusted more than about half a turn, it is recommended that the whole procedure be repeated, starting from the 1650 kHz setting.

If there is a fault in the circuit, this will often manifest itself as the RF tuning being difficult to achieve and items that are adjusted appear to be at the limit of their range. Ideally, the adjustment rods of the ferrite cores in the tuning mechanism should be all about the same protrusion from the front panel metalwork when positioned at the high end of the frequency range.

10.7.8 Replacement Parts

The Astor car radio uses many parts no longer manufactured and so finding modern equivalents can be difficult. This is particularly so for Germanium semiconductors of the PNP type. Table 10.7.2 provides some suggestions.

Part	Description
Front panel globes	BA7s Indicator Light, Clear, 12 V, 100 mA, 5000h
RF Amplifier transistor	PN3693 (2N3693) Fairchild Transistor (TO-92)
Osc, Mixer, IF transistors	BC546A, BC548 NPN
Audio volume transistor	BC149, BC109 NPN
Driver transistors	2N6520TA PNP Transistor, 500 mA, 350 V, 3-Pin TO-92, BC327
Audio power transistors	AT1138, AD149, 2SB337, 2N301, OC26 PNP Germanium
Power line iron core choke	128 turns 0.5mm dia copper wire
Speaker transformer	2 x 130 turns centre tap primary, 220 turns secondary
Bias current trim pot	220Ω, Through Hole Trimmer Potentiometer 0.5W Top Adjust Bourns, 3329
Temperature compensation	ASC125 2N408 PNP Germanium

Table 10.7.2 *Replacement parts.*

10.7 Radio Interference Suppression.

There are many sources of radio interference to be found in older model cars. The worst offenders are the high voltage spark plug leads, in former days, made from copper wire. Other sources of noise are the generator brushes, wiper motor brushes, and sometimes even static electricity generated at the tyres.

All these sources have one thing in common – a spark. It was H. Hertz who discovered that a spark can be the source of electromagnetic waves. Indeed, this is how Hertz discovered the existence of electromagnetic waves – light at that time not being recognised as such.

With a spark, not only is light emitted, but a whole spectrum of electromagnetic waves and since these waves fall into the radio spectrum, are received by the radio aerial and amplified along with the desired station.

It is therefore common practice to eliminate this source of radio interference by the use of a suppressor in the ignition leads.

It is the rapid fluctuations in current across the plug gap result in emission of radio frequency electromagnetic radiation (i.e., static). The "time constant" of such a circuit depends on the product of the resistance R and the capacitance C. Increasing the resistance of the circuit increases the time constant, and so the fluctuations are at a lower frequency – leading to reduced radiation emission in the radio spectrum, thus reducing the static that we hear.

In the case of the ignition leads, the circuit consists of a resistor (the wire) and a capacitor (the plug gap and the distributor cap gap).

10.8 Wiring and Fuses

10.8.1 Wiring Harness

Wiring harnesses for vehicles using Lucas electrical equipment follows a standardised design. Most wires for low current systems are specified as either 9/0.012 (5A capacity) or 14/0.012 (7A capacity) with rubber and fabric insulation. Heavy duty cables such as those for headlamps are 28/0.012 (14A capacity) while the main battery feed from the starter switch and generator output is 44/0.012 (22A). The first number in these specifications being the number of strands and the second, the strand diameter in inches. 0.012" diameter cable is equivalent to 30 SWG. Heavy copper stranded cable 37/20 SWG is used for the battery connection and starter motor. In general, but dependent on conditions, the current carrying capacity of the wires in the harness are:

Wire size	Current Carrying Capacity
9/0.012	5
14/0.012	7
28/0.012	14
44/0.012	22
37/20 SWG	Starter Cable 400A

Table 10.8.1 *Current carrying capacity of wire.*

Feed wires are brown for current supply or feed to all services, white for ignition-switch controlled circuits, yellow for generator and field circuits.

Colour	Application
Red	Side and tail lamp circuits starting from light switch.
Blue	Headlamp lighting circuit starting from light switch.
Purple	Auxiliary circuits protected by a fuse A2.
Green	Auxiliary circuits fed through the ignition switch and protected by fuse A4.
Black	Earth circuit.

Table 10.8.2 *Lucas circuit colour coding.*

Feed wires have the main basic colour only. Switch wires have the main colour of the feed with a coloured tracer. All earth wires are black.

When diagnosing electrical faults in a vehicle's wiring system, it should be realised that voltage drops only occur when current flows.

For example, consider a case where there is a problem with say the fuel gauge sender unit. An operator may remove the wire from the sender unit terminal and measure 12V, and so assume that the sender unit is receiving a full +12V. This is an open-circuit voltage. However, if there is a high resistance upstream of this connector, such as a burned out wire, excessive crimp, or corroded contact somewhere in the wiring loom, then when the wire is reattached to the sender unit, there may be a considerable voltage drop across the fault due to current now flowing. The actual voltage applied to the sender unit be much lower than expected. Often, it is not possible to measure the voltage at a terminal when it is connected because of the insulating covers in place. Measuring the "open circuit" voltage may not necessarily indicate where the problem is even if +12V is measured in that condition.

10.8.2 Fuses

Fuses offer protection against electrical overload and possible fire. A fuse is a short piece of wire enclosed in a glass tube with metal cylindrical caps on each end. The wire is designed to melt when carrying a current in excess of its rating, so breaking the current in the electrical circuit to which it is connected. Fuses are of course connected in series in the circuit that they protect.

There are two rating specifications to consider, the most important being the current rating. In automotive use, this usually ranges from 1 to about 35A.

The fuse melts because it dissipates power as heat via its internal resistance R. The power dissipated P depends upon the square of the current I.

$$P = I^2R$$

If the fuse cannot dissipate electrical energy at a rate at which it is delivered, its temperature rises and the fuse melts. It follows then that a fuse with a larger current rating has a lower resistance so as to be able to pass more current before melting. It is sometimes mistakenly believed that a 12V fuse at say 10A would be dissipating 120W. This is not correct. The voltage rating is not the voltage across the fuse. The voltage across the fuse is very small. It is the current rating that is important as far as power dissipation, and hence heating, is concerned. The voltage rating is more important for high voltage circuits such as in mains-operated equipment. At higher voltages, the fuse should not have the capability of arcing to a nearby component, especially when melted and the two remaining ends of the fuse may be free to move. The voltage rating is the voltage from the fuse to "ground", not the voltage that is dropped across the fuse.

Unfused circuits are usually:

- Alternator/Generator
- Headlamps, panel lights, front and rear tail lights
- Ignition coil
- Fuel Pump

Fused circuits are:

- Horn
- Wiper motor
- Fuel gauge
- Stop lights, Indicator lights.

There are several styles of fuses in general use.

Fuse Type	**Vehicle origin**	**Dimension**
3AG	American	1 ¼"L x ¾"D
8AG	American	1"L x ¼"D
ENG	English	7/8"L to 1 5/32"L x ¼"D
SFE	American	5/8"L to 1 7/16"L x ¼"D
CERAMIC	Continental	15/16" x ¼"D

Table 10.8.3 *Fuse types.*

Note that American fuses are longer than English fuses.

CHAPTER 11. BODY

11.1 Body Engineering

11.1.1 Introduction

The design of a motor vehicle body as distinct from the mechanicals is of course of utmost importance in determining the success of the vehicle in the market. There are two aspects to body design which should, in ideal circumstances, work hand in hand: body engineering and styling.

The most critical aspect of body engineering is mass, or weight. Overall weight is a determining factor for the cost, and hence selling price of the model. For vehicles manufactured and sold by BMC in Australia, body design was largely outside the control of local engineers – it being done in UK. However, during the years 1968 to 1973, the local company had the opportunity to design and build a vehicle model range entirely from scratch – Model A, and Model B. Model B was to become the Leyland P76. Model A was eventually named P82, but never made it beyond a prototype stage. As such, local engineers took on a significantly greater responsibility than previously. The only design target at this point was that the factory should produce a large vehicle to compete with other local manufacturers who were enjoying considerable sales successes with their models. As well, to cater for the market for smaller vehicles, that within which BMC were more familiar, a new small to medium sized model would be designed with many common design features as the larger car.

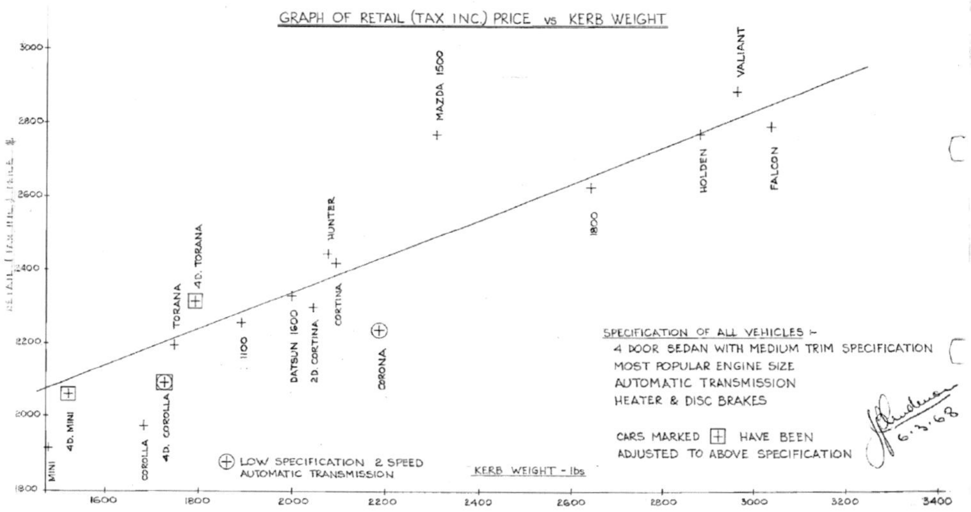

Fig. 11.1.1 *Retail price vs kerb weight for various vehicles in the Australian market both large and small, 1968.*

In 1968, Anderson and Hardy[31] reported that in Australia, 4 door automatic transmission sedans, fitted with heaters and disc brakes, retail at $1.06 per lb kerb weight. Fig. 11.1.1 shows the results of their survey of the field.

Hardy and Anderson took the percentage weight allocations of vehicle components used by manufacturers of conventional cars (north south engine, rear wheel drive) and adjusted them according to the characteristics of BMC's front wheel drive range, to arrive at a point where potential weight reductions could be made.

Any saving in costs resulting from a weight reduction, such as use of a light alloy engine block, would have to be balanced against any increase in manufacturing costs.

Increases or decreases in the weight of a component has a follow-on effect in terms of parasitic weight associated with that component. For example, a larger engine will require heavier front suspension, wheels, engine mounts, radiator, fuel tank, etc. Generally speaking, any increase in mass of one component requires a 50% penalty for parasitic mass.

As discussed previously, it was shown that mass savings have a direct effect on fuel consumption according to[32]:

$$L/100km \sim Mass^{0.4} + Eng.Cap.^{0.25}$$

And acceleration times as:

$$Acceleration\ time \sim \sqrt{Mass} \times \sqrt{Eng.Cap.}$$

Motor vehicle developments since the 1970s have been remarkable in terms of the attention given to this most important issue – from the use of high strength steels (thus allowing thinner panels to be used), to the use of plastics and light weight alloys in body and mechanical systems. It is all the more remarkable that such advances have occurred despite the huge number of systems now expected as standard on a modern motor vehicle, such as air conditioning, power steering, driving aids, not to mention prescribed increases in safety and lower emissions.

The body engineer is not only concerned with body sheet metal. In the Schedule of Parts, Groups A and B shows that the comfortable accommodation of passengers extends to many other areas of responsibility:

Group A Body Shell	
A1	Body in White
A1-A	Index to Schedule
A1-B	Body Shell Less Doors
A1-C	Front Panel and Front Wings
A1-D	Bonnet
A1-E	Doors – front and Rear
A1-F	Trunk Lid or Tail Door
A1-G	Loose Parts

31 G.B. Hardy and J.B. Anderson, "Product Studies (Passenger Vehicles) for 1973-1978", BMC Australia, 1968.

32 R. Habgood

Group B Body Fittings and Trim	
B1	Bonnet Fittings
B2	Door Fittings
B3	Trunk Fittings
B4	Front Seat and Trim
B5	Rear Seat and Trim
B6	Facia and Dash
B7	Windscreen Fittings
B8	Backlight Fittings
B9	Floor Fitting and Covering
B10	Interior Lining
B12	Miscellaneous
B21	Radiator Grille
B22	Motif and Décor
B23	Bumpers Front and Rear

Table 11.1.1 *Sub-groups of Groups A and B of a vehicle Schedule of Parts showing breakdown of body components.*

In this chapter, we are not so concerned with weight savings associated with the mechanical design, but more, the body shell. Market acceptance of a new model depends largely on those items seen and appreciated by the customer in the showroom. To this end, styling, structural integrity, and accommodation of passengers and luggage are of prime importance.

11.1.2 Volume

Body engineering is concerned with maximising interior volume while minimising sheet metal and at the same time, producing a vehicle of an acceptable strength and appearance.

In its most basic form, the purpose of a passenger vehicle is to convey passengers from one place to another at minimum cost, minimum time, and maximum comfort. We've already given some attention to vehicle performance and suspension, and so here we are concerned with volume. That is, the volume within which the passengers and their luggage are to be accommodated with the minimum of sheet metal.

As a simple exercise, we can imagine a motor vehicle as an object such as a rectangular prism.

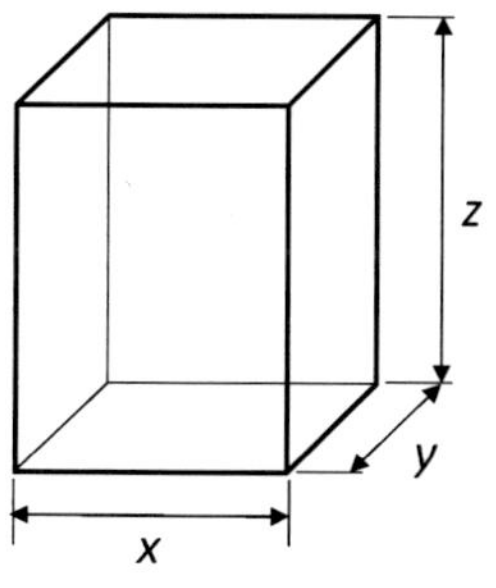

Fig. 11.1.2 *A simple body in the form of a prism.*

The volume is found from:

$$V = xyx$$

The surface area is:

$$S = 2xy + 2yz + 2xz$$

Our aim is to determine the minimum surface area *S* for a particular volume *V*. The first step is to express *S* as a function of *V* and to consolidate the number of variables. Thus:

$$S = 2xy + 2\frac{V}{x} + 2\frac{V}{y}$$

To find the minimum value of *S*, the partial derivates of *S* with respect to *x*, and *S* with respect to *y*, are set to zero. In effect, we hold one dimension constant and find the value *x* that minimises *S*, and then hold the other dimension constant and find the value of the *y* that minimises *S*. The third dimension *z* is tied to *x* and *y* via its relationship to *V*.

$$\frac{\partial S}{\partial x} = 2y - 2Vx^{-2} = 0$$
$$\frac{\partial S}{\partial y} = 2x - 2Vy^{-2} = 0$$

This means that:

$$y = \frac{V}{x^2}$$
$$x = \frac{V}{y^2}$$

or,

$$x^2y = y^2x$$
$$xy = y^2$$
$$x = y$$

Thus, one condition for minimum surface area with respect to volume is that two sides have equal length.

Now, the height z is given by:

$$z = \frac{V}{xy} = \frac{V}{x^2}$$

Going back to S, we have (with *x* = *y*):

$$S = 2x^2 + 4xz = 2x^2 + 4\frac{V}{x}$$

To minimum *S* with respect to *x*, the derivative of *S* with respect to *x* is set to zero:

$$\frac{dS}{dx} = 4x - 4\frac{V}{x^2} = 0$$

$$x = \frac{V}{x^2}$$
$$V = x^3$$

That is, for a given volume *V*, to accommodate our passengers, the dimensions of the box should be cubic, with *x* = *y* = *z*.

It is therefore no accident that the admirable volume to surface area ratio of a Morris Mini has a box like shape approaching that of a cube, so much so that the car is often called a "brick". A Leyland P76 boot has a very large capacity compared to its rivals, not only because it looks big, but is also somewhat "cubic" in aspect ratio.

But of course, human beings are not cubes, and their height and width have to be accommodated, let alone more volume for the engine, rear axle, and boot. One of the very first tasks of the design of the body for the Leyland P76 was to construct a seating buck, in which space for power plant and passengers could be arranged as shown in Fig. 11.1.3.

Fig. 11.1.3 *Seating buck for Model B (Leyland P76), 1968.*

From this, then all else follows.

However, the arrangement for seating is not a haphazard one. In 1968, Anderson and Hardy knew that there would be limitations set by the upcoming Australian Design Rules in 1970 whereby passengers of the size from 5% of female, and 95% of male passengers, would require minimum clearances and visibility.

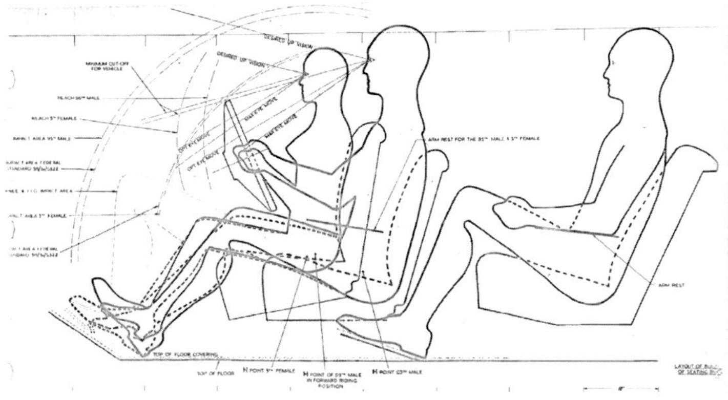

Fig. 11.1.4 *ADR requirements for passenger accommodation.*

These restrictions, together with mechanical considerations, would set design hard points in dimensions which would constrain significantly the work of the stylist. It was for good reason that the local stylist Romand Rodbergh woefully observed that all the styling submissions produced by him and Michelotti looked very similar.

Given the volumes required for passengers, luggage and mechanicals, the overall dimensions of the body are thus set. The stylist produces an outline shape, and then the body engineer sets about how to make this shape in the most economical manner. Not only are the body panels to be optimised, but also the tooling that is to make them.

In the case of the Leyland P76, an innovative approach to these problems was taken by reducing as much as possible the number of panels that had to be welded or otherwise joined together. Every joint costs money in terms of labour, equipment and weight.

11.1.3 Strength

BMC Australia went to some lengths to test vehicles for suitability for local conditions – which were invariably much harsher than those experienced in UK. Body strength and reliability testing, both laboratory and field, became even more important with the introduction of mandatory safety features introduced through the Australia Design Rule system.

Fig. 11.1.5 *Seat belt anchorage testing Morris Mini Deluxe YDO5.*

Local modifications to body work were limited to the addition of braces and stiffening panels where needed.

For the Leyland P76 body, significant engineering at a more fundamental level was required. Although the chassis engineering for the Leyland P76 was performed at Product Engineering in Australia, body engineering was the responsibility of Press Steel Fisher in UK. Pressed Steel Fisher describe the P76 saloon body thus:

"This is a 4 door body of unitary construction for manufacture by BLMC Australia. The power unit is mounted conventionally north-south at the front and transmission is by a single drive shaft to the rear wheels. Suspension is by Macpherson strut at the front and trailing arms with coil springs at the rear. Windshield and backlight glasses are mastic glazed. Windshield wipers park below the rear edge of the bonnet which forms a gap with the glass. The spare wheel is housed in the LH side of the luggage compartment in a vertical position. Die cast extensions are bolted to the front and rear fenders. The body assembly is by the "Gate Line" method. Provision is made for the Roto-Dip paint process. Front sidemembers of the underbody are welded to the bottom of the valance and extend rearwards beneath the floor but are not attached to it. Connection to the A post is made via a torsion box. Engine and gearbox are mounted on 2 crossmembers bolted to the front and rear of the sidemembers. The one floor crossmember is located at the rear and provides for the rear shock absorber mounting. The dash panel extends from the toeboard to include the facia top and forms the glazing surface for the lower windshield. Three roof bows of light section are bonded by adhesive. The header panels and cantrail closing are similarly attached. In order to accommodate the petrol filler pipe completely within the wheel arch, the wheelarches are higher than normal. The petrol filler area of the wheelarch is bonded to the outer panel with adhesive. The squab panel is bolted into the body full trimmed and is identical to the folding squab used on derivatives. A full inner panel for the bonnet is employed with clinched edges. Front hinges are used with a single central lock at the rear. Torsion bars assist in opening with a stay for security. The full inner panel is used for the trunk lid with clinched edges. The hinges are welded to the body prior to assembling the torsion bars which act directly through the centre line of the hinge to balance the lid. The claw type lock has common parts with the bonnet lock."

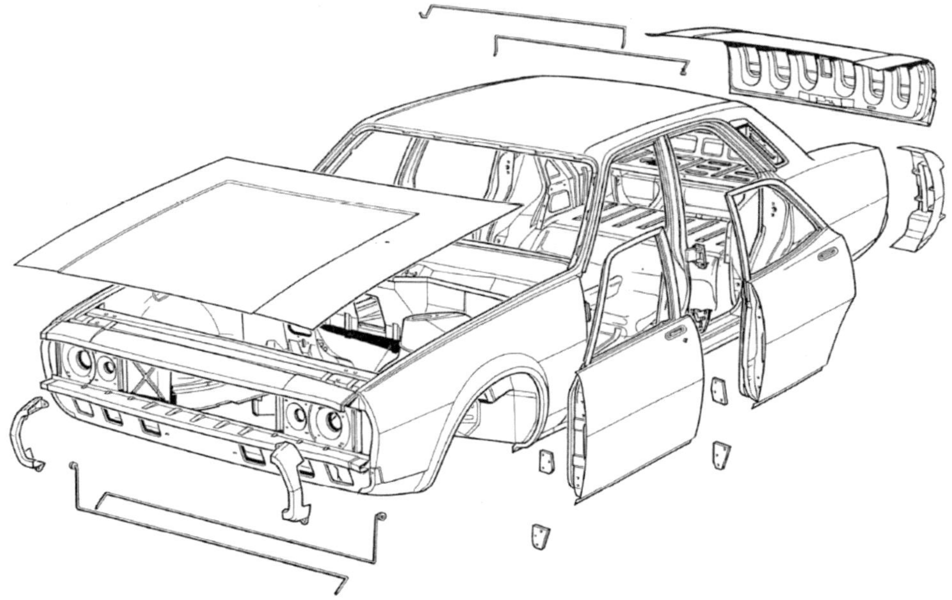

Fig. 11.1.6 *Pressed Steel Fisher body engineering report, Model B (Leyland P76) 1971.*

11.2 Styling[33]

Styling at BMC Australia was historically limited to the addition of chrome strips and other embellishments on what were basically UK designs. Despite this, the Australian stylists did have some notable successes in lifting the sometimes drab UK designs to more appealing creations (even though it was found that one member of the powerful styling committee was colour blind), via interior and exterior colours, and re-designed grilles and headlamp assemblies.

Early on, Managing Director Buckley foresaw the time when BMC Australia would need to make vehicles more attuned to Australian conditions, not only mechanically, but also stylistically, and formed a styling department within Product Engineering in 1958.

The company's first stylist, Bill Moody, transformed the Austin A95 Westminster into the Morris Marshal with a stylised boomerang in the front grille and a dummy air intake in the bonnet.

Fig. 11.2.1 *Morris Marshal, 1958.*

When development began on the Leyland P76 in 1968, the issue of motor vehicle styling took on a new importance. The approach was to follow the Ford "Theme Car" system which divides the development of a motor vehicle into five stages and from which timing network diagrams could be constructed to cover engineering, tooling, manufacturing and marketing operations.

At Ford, the styling department was typically concerned with establishing the critical marketing characteristics of future vehicle concepts backed by a complete list of product assumptions based on all the known competitive and marketing information.

In those days, preparation of a full-size clay or fibre glass model provided a first look at any new vehicle concept at low cost and compared with projected future competitive vehicles illustrated by full-size renderings or tape drawings. The stylist did not have a free hand. Not only were market forces to be considered, but more importantly, the engineering hard-points which had to be adhered to, and often, to less than millimetre. Typical hard-points are:

33 Acknowledgement David Bentley

224 THE BMC ENGINEERING COMPANION

Seating package
Body cross section in the greenhouse area
Side door openings
Centre pillar including glass channels
Hinge lines
Glass panes
Inner face of doors
Rocker panel
A pillar

Windshield surface
Cowl top surface
Dash panel
Underbody assembly
Inner rear wheel arches
Steering wheel arches
Steering wheel location
Spare wheel location

Although the intention was that Engineering and Styling would work hand-in-hand to produce the desired vehicle, at Leyland Australia, the engineers dominated the process and the stylist left to make what he could of the remainder.

The Design Terms of Reference of 1969 stated:

> *"The styling of Model 'B' range is critical. It must appeal to customers who are used to American styling but must have a character of its own. Furthermore, in some versions, it must appeal to our faithful but rather conservative customers who are used to English products."*

Despite the sentiment above, the Styling Terms of Reference really amounted to a long list of engineering requirements that the stylist would have to satisfy. Styling for the forward portion and windshield side were to be common between sedan, wagon and sports coupe. Even the bonnet style would be dependent on the diameter of the V8 engine cooling fan. For the front grille, one of the most important and visible aspects of a vehicle's personality, it was deemed that:

> *"the styling of this component must be aimed at minimising the expenditure. Preference is for an unobtrusive mesh placed immediately in front of the radiator and condenser cores."*

In Wheels Magazine of March 1973, Managing Director Peter North is quoted as saying that the P76 "was styled after the car had been designed."

Fig. 11.2.3 *The Leyland P76 Wagon, Saloon and Coupe.*

What lies behind the somewhat nebulous concept of styling was best elucidated by David Bentley, a trainee stylist at Leyland Australia during the early stages of the Model B, P76, program. Bentley writes:

> *"First impressions do count, and in most cases, the first impression for a motor vehicle is a visual one. This precise moment, when a person sees a product for the first time, initiates a very complex association between product and person."*

Some cars leave a "total" impression while others register some particular aspect and leave a "feature" impression. For example, Bentley asks us to consider the difference between a Mercedes (total impression: slick elegance), or a Holden Torana XU1 (total impression: flashy and aggressive). On the other hand, the huge rear fenders of a two-door Ford Landau, or the large curved rear window of a Jensen Interceptor, are feature impressions that register at once with the viewer.

Bentley says:

> *"The reason some cars impart total impressions while others leave feature impressions results from the degree of visual harmony achieved. Visual harmony means there is an aesthetic association between the various visual elements of the car in the same way that there is an engineering association between mechanical elements: just as there is a relationship between the engine performance and suspension so is there one between the shape of the front fender and that of the rear lamps.*
>
> *A mechanical mismatch becomes apparent by feel when driving, an aesthetic fault is apparent visually. Good mechanical association results in a feeling of unity, and good aesthetic association produces visual harmony.*
>
> *Cars which leave total impressions mostly have a look of professionalism. While professional styling strives for aesthetic harmony without fashion, commercial styling tends to display pronounced individual features at the expense of overall harmony.*
>
> *As with any art form – and styling is an art form – beauty is in the eye of the beholder. But when fashion is present in a particular art form its field of public acceptance narrows. This is because in promoting the goods, you are saying to the public "here is my interpretation of a particular fashion trend, I hope it coincides with yours so that you will buy my goods." Thus, those people who don't happen to like the fashion, may be eliminated as potential customers.*
>
> *A similar philosophy can be applied to cars which have aesthetic harmony and those which exhibit some distinctive feature. When people see a car because of some specific feature, and not because of its total appearance, they react toward or against the whole car in terms of that feature. They may not, in fact, even see the feature but they feel there is something about the whole car which just doesn't seem to be right. What styling has done then is to create a feeling of uncertainty in the mind of the potential customer; an uncertainty which may cause him to turn to another car which he feels more at ease. This is why it is more important to achieve a high degree of aesthetic harmony in commercially styled cars."*

The present author feels the above is a most profound statement and applies to many areas of industrial design besides motor cars.

The customer who buys a particular vehicle knows that they will be seen in public with their purchase, whether it be on the road, or parked in their driveway. Indeed, the purchase of a

226 THE BMC ENGINEERING COMPANION

particular motor vehicle is a public statement of support for the vehicle's style, and to a lesser extent, its engineering. Should the customer not wish to be personally associated with a vehicle, it can always be said that "it's a company car", but for the most part, the car is an extension of the customer's personality. a statement of his or her astuteness, and an indication of aesthetic taste.

Styling at BMC/Leyland both in UK and Australia always appeared to have an uneasy relationship with engineering. When leaving the UK company, Roy Haynes wrote to Australian Managing Director David Beech saying

> *"It all gets back to how much importance you attach to the value of styling and what sort of car is eventually reflected to the buyer."*

11.3 Rust

11.3.1 Chemical Reaction

In simple terms, the complex redox reaction for the rusting of iron and steel is written as:

$$4Fe(s) + 3O_2(g) \rightarrow 2Fe_2O_3(s)$$

In practice, the rusting of iron requires both air (O_2) and water, and the reaction proceeds as a series of steps. The oxidation half-reaction is:

$$Fe(s) \rightarrow Fe_2^{2+} + 2e^-$$

And the reduction half-reaction is:

$$O_2 + 2H_2O + 4e^- \rightarrow 4OH^-$$

The OH^- ions react with the Fe^{2+} ions:

$$Fe^{2+} + 2OH^- \rightarrow Fe(OH)_2$$
$$4Fe(OH)_2 + O_2 + 2H_2O \rightarrow Fe(OH)_3(s)$$
$$Fe(OH)_3(s) \rightarrow Fe_2O_3, H_2O + 2H_2O$$

The brown coloured metal we usually associate with rust is the compound

$$Fe_2O_3, H_2O$$

an Fe(III) hydrated oxide.

Iron and steel often rust badly because the rust has a lower density than the parent metal and tends to flake off, thus exposing more of the material to water and oxygen and promoting further rusting.

Water and oxygen are typically excluded by painting the article. However, a very effective way to reduce rusting of iron is to attach a more reactive metal (i.e. one with a lower E^o so that the iron undergoes reduction and the sacrificial anode (usually zinc) undergoes oxidation. This process is often called cathodic protection. The chemical species involved are:

$$Fe / Fe_2^{2+}$$
$$E^{o} = -0.44V$$

And

$$Zn^{2+} / Zn$$
$$E^{o} = -0.76V$$

Iron and steel items protected in this way are said to be "galvanised".

Aluminium undergoes corrosion in the presence of atmospheric oxygen. A very thin (nm) tough layer of Al_2O_3 (alumina) appears within nanoseconds on the metallic surface which shields the parent metal from further corrosion. Alumina is a very hard ceramic material and is a non-conductor of electricity.

11.3.2 Rust and Corrosion

11.3.2.1 Rust in steel panels and pipes

The front section of an exhaust pipe in a motor vehicle is seldom observed to be rusted out because it usually gets hot enough to drive off the water in the exhaust gases before serious rust can take hold. It is the rear sections (usually the rear muffler), that remain relatively cool that are first to go due to condensation of water on the inside.

However, what is curious is that the mufflers and tailpipes, which usually rust out from within, generally have a non-rusty appearance from the outside whereas the front section usually appears very rusty from the outside but remains relatively sound for many years.

The front section of the exhaust, particularly the exhaust manifold, gets very hot compared to the remainder of the exhaust system. Rust is the combination of iron, water and oxygen. As with many chemical reactions, there is usually an energy barrier to be overcome before the reaction proceeds. In the case of rust, the water and oxygen molecules have to prise out an iron atom from the metal before iron oxide is formed. This takes energy. When the metal is hot, its atoms are jiggling around much more violently compared to when the metal is cool. Thus, hot metal rusts more easily than cool metal because the iron atoms in the hot metal are already somewhat loosened by their thermal motion and so the water and oxygen molecules find these atoms easier to combine with to form rust.

But close to the engine, this very heat drives off the water needed to form rust and so only light, but very visible, rust occurs. The muffler and tailpipe generally do not show signs of rust from the outside but rust heavily from within due to the condensed (and sometimes stagnant) water present.

Much the same effect happens when metals are bent or otherwise distorted. For example, it is found that rust begins more easily on creases in steel body work, or where there has been some welding because of the deformation of the metal causes atoms to be displaced from their usual positions and become less well bonded – and so are more amenable to being captured by water and oxygen to form rust. Heating and bending should therefore be avoided as much as possible.

11.3.2.2 Corrosion in alloy parts

The process of corrosion is generally an oxidation-reduction reaction in which the species with the lower value of standard potential gives up electrons to that with the high potential. Such a reaction requires the presence of two dissimilar metals. The process is aggravated by heat for reasons given previously.

In the case of alloy parts on an engine, it is found that a part like a thermostat housing is prone to corrosion because of its attachment to the cast iron cylinder head, and the water flowing through both acts as an electrolyte.

Since the aluminium alloy corrodes preferentially to iron, material loss occurs where it is in contact with the iron. This is usually found to be in the area of the fixing stud holes and the flat face abutting the thermostat.

CHAPTER 12.
LIGHTING & MIRRORS

12.1 Introduction

Most repair and parts books put lighting in with the electrical system, but there is more to motor vehicle lighting than just the way in which they are powered. In the early days of the motor vehicle, lights were power by burning gases. Once a source of light is established, whether it be by heat, or some other means, then the process of light transmission to outside the vehicle is important.

As well, mirrors both internal and external, are important safety features of any motor vehicle and the passage of light incident and reflected from them deserves some attention and analysis.

These topics are therefore the subject of this chapter.

12.2. Reflection and Refraction

12.2.1 Reflection

Consider a light source in the form of a point. The directions in which light waves radiate from the source are indicated by straight lines called rays. Rays are drawn perpendicular to the wave front of the light waves.

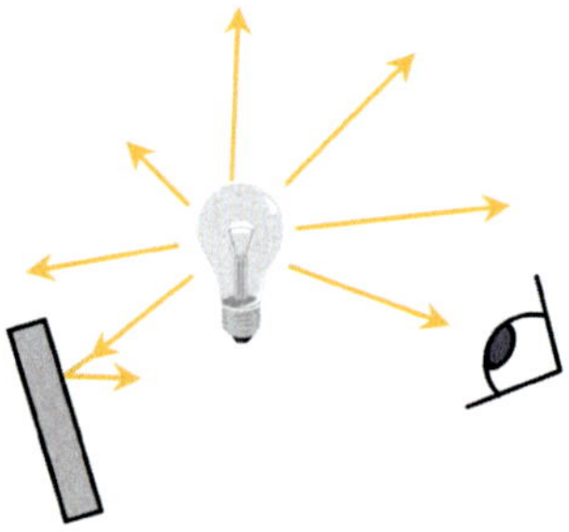

Fig. 12.2.1 *Light rays emanating from a light source.*

Rays of light emanate from the source and radiate outwards in all directions. Some rays enter an observer's eye and are focussed by the lens in the eye to form an image of the source on the retina. Other rays are reflected and/or absorbed by the surroundings.

When light rays strike a surface, the angle of reflection = the angle of incidence.

$$\theta_i = \theta_r$$

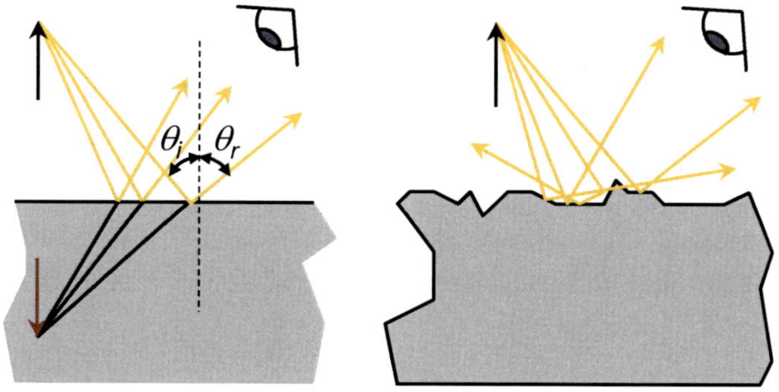

Fig. 12.2.2 *Reflection from smooth and rough surfaces.*

If the surface is smooth, then the reflected rays may enter the eye of an observer who sees an image of the object as if it were positioned behind the surface.

If the surface is rough, the angle of reflection still equals the angle of incidence for each ray, but these angles are now such that the light is scattered. Many of the light rays do not enter the eye at all and an image cannot be formed. Or, the ones that do enter the eye appear not to come from a single point, so again, no clear image is formed.

12.2.2 Refraction

If a light wave strikes a medium (such as a block of glass) at an angle, then the waves which enter the medium slow down and the wavelength becomes shorter to compensate. This has the effect of altering the direction of travel of the wave.

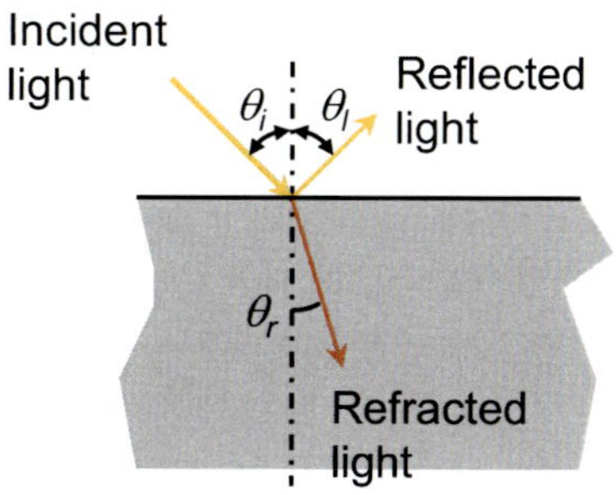

Fig. 12.2.3 *Refraction in a medium.*

Experiments show that for light of a single wavelength, the angles of incidence and refraction are related by the refractive indices *n* of the two materials:

$$n_2 sin\theta_1 = n_2 sin\theta_2$$

The refractive index of glass is about 1.52, and of air, about 1.003.

If light travels from a material with a low value of *n* to one with a high value of *n*, then the velocity of the wave is reduced and the wavelength becomes shorter to compensate. The frequency remains the same. The path taken by light rays is reversible. It doesn't matter whether the light rays travel from an optically more dense to less dense material or vice versa.

12.2.3 Mirrors

Of the many rays of light that might strike a mirror, the path of two of them is known no matter what kind of mirror or object is involved. The two rays are:

1. Any incident rays which are parallel to the principal axis are reflected such that the reflected rays appear to come from the focal point *f*.
2. All rays passing through (or travelling towards) the centre of curvature *c* of the mirror strike the mirror at right angles and hence are reflected back along their original path.

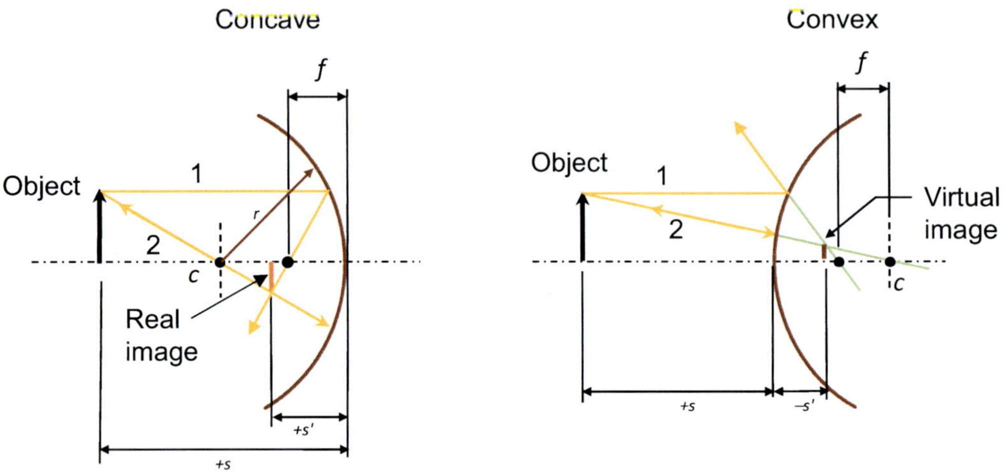

Fig. 12.2.4 *Rays incident and reflecting from a mirror.*

There are two sides to a mirror. The side that receives the incoming light is called the real side. The side where there are no light rays present is called the virtual side.

The eye interprets the light rays emanating from a point in space as coming from an object. The image formed by the mirror is the object for the lens of the eye. The eye "sees" the image of the actual object in the mirror. The image may be magnified or diminished in size, inverted or upright in orientation, depending on where the object is situated with respect to the mirror and the curvature of the mirror. The image might be located on the real side or the virtual side of the mirror.

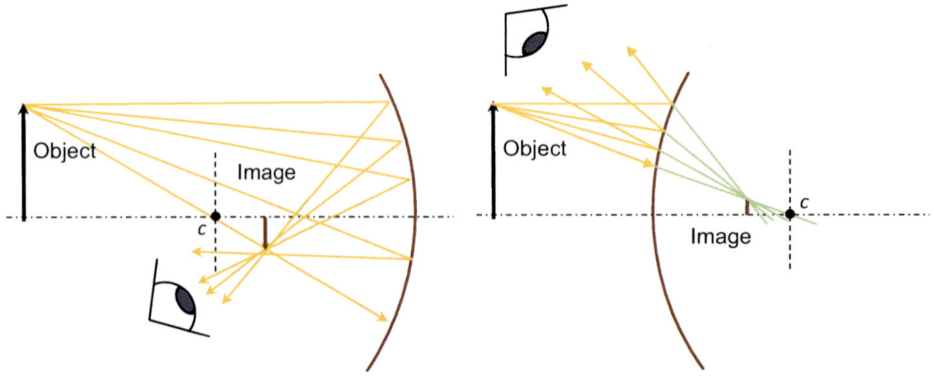

Fig. 12.2.5 *Reflected image seen by the eye for a concave and convex mirror.*

The object distance, image distance, focal length and radius of curvature are related by:

$$\frac{1}{s}+\frac{1}{s'}=\frac{2}{r}=\frac{1}{f}$$

In this formula, *r* is the radius of the mirror, and *s* and *s'* are the object and image distances from the front of the mirror, *r* is the radius of the mirror and f is the focal length. The focal length *f* is positive if the centre of curvature lies on the real side and negative if the centre of curvature lies on the virtual side. The focal length *f* has the same sign as *r*.

The linear magnification *m* is the ratio of the image size to the object size:

$$m=\frac{h'}{h}=-\frac{s'}{s}$$

The side mirror of a modern motor vehicle is often inscribed with the words, "objects in the mirror are closer than they appear" because the magnification is reduced, at the expense of a wider field of view. These mirrors have a negative (convex) curvature.

12.2.4 Lenses

A lens consists of two refracting surfaces. Here we consider two spherical surfaces which are positioned close together – a thin lens. The general convention, for objects or images, is that "real" is positive. Note that unlike a mirror, the virtual side of a lens for images is the real side for objects.

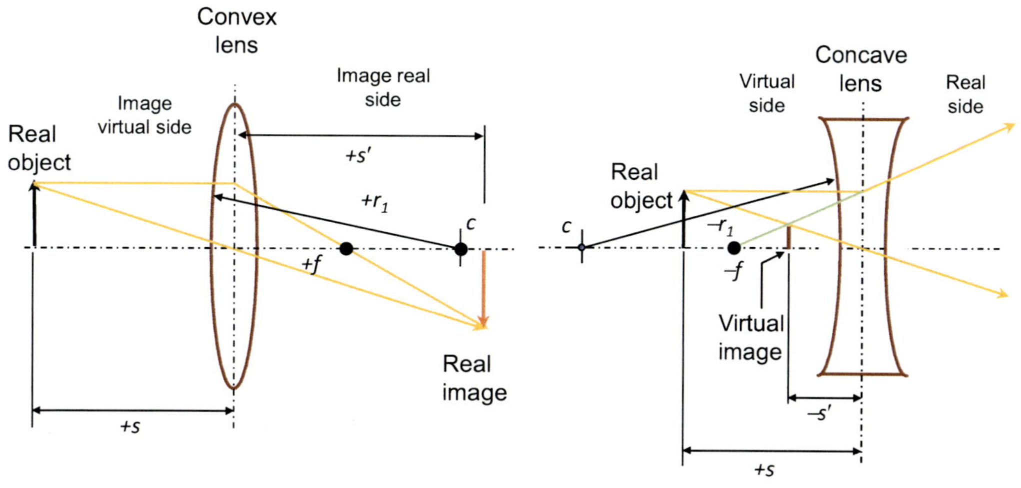

Fig. 12.2.6 *Passage of light through convex and concave lenses.*

For light rays travelling from left to right through a lens, an image is "real" if it is formed on the right hand side of the lens and "virtual" if formed on the left-hand side. These terms are just labels for identifying on which side of the lens the image or objects are. If the rays diverge from an object, the object is real and is on the left side of the lens. If the rays (going from left to right) converge to the object, then the object is virtual and lies on the right side of the lens.

The image shown for the concave lens in Fig. 12.2.6 is a virtual image since it is formed on the left-hand side of the lens and thus the image distance s' is negative. The radius of curvature is +ve if it lies on the real side. In all formulas, r_1 is the radius of curvature of the first surface upon which light is incident.

For a lens, the focal length f is positive if the incident parallel light converges to form a real image (e.g., convex). Focal length is negative if incident parallel light diverges to form a virtual image (e.g., concave).

The path of two light rays through a thin lens system is known no matter what kind of lens or object is involved. The two rays are:

1. A ray that passes through the centre of a lens continues on in a straight line.
2. A ray that travels parallel to the principal axis will emerge from the lens in a direction towards the focal point (and vice versa).

Compared to a mirror, a lens has two focal points, one on each side of the lens and two radii of curvature. r_1 is the radius of curvature of the first surface upon which light is incident. The focal length is *not* one half of the radius of curvature for a lens. The two focal lengths for a lens are equal, even if the radii of curvature for each side are different (if the medium is the same on each side of the lens, e.g., air).

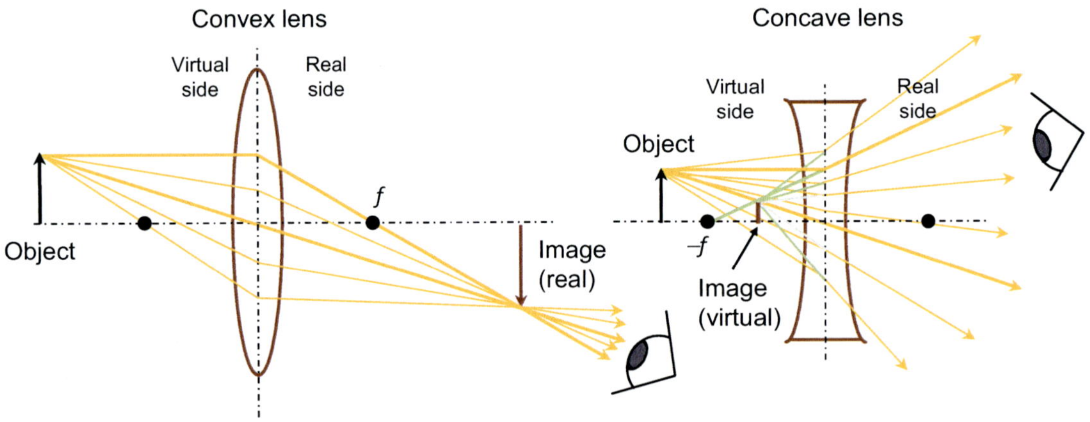

Fig. 12.2.7 *Images formed by a lens.*

The eye interprets the light rays emanating from a point in space as coming from an "object." The image formed by the lens is the object for the lens of the eye. Compared to the actual object, the image may be magnified or diminished in size, inverted or upright in orientation, depending on where the object is situated with respect to the lens and the nature of the lens.

The object distance, image distance, focal length and radius of curvature are related by the thin lens and lens maker's equations. The thin lens equation is:

$$\frac{1}{s} + \frac{1}{s'} = \frac{1}{f}$$

And the lens maker's equation is:

$$\frac{1}{f} = \left[\frac{n_2}{n_1} - 1\right]\left[\frac{1}{r'} - \frac{1}{r}\right]$$

The linear magnification m is the ratio of the image size to the object size:

$$m = \frac{h'}{h} = -\frac{s'}{s}$$

where n_1 is the refractive index of the medium (e.g. air), and n_2 is the refractive index of the lens material.

The power of a thin lens in air is given by the reciprocal of the focal length and has units dioptres.

$$P = \frac{1}{f}$$

12.3. Headlamps

Headlamps are used primarily to light the road ahead so that the driver may see where he or she is going at night. There are two problems to be overcome. First, the source of the light must provide as white a light as possible, consume as little current as possible, and withstand the mechanical and electrical shock of going from room temperature to full heat in less than a millisecond. Secondly, once the light is produced, it must be directed to illuminate the road ahead without dazzling oncoming drivers.

As far as the headlamps go, the light source for BMC cars of the 1960s is a filament globe located at the focus of a parabolic mirror. Rays which reflect off the mirror are parallel. They then pass through the front lens of the headlamp and are refracted. A shield is placed over the filament to block direct transmission from the filament to the glass lens. Sometimes the shield takes the form of what is called a "getter", a silvery film inside the bulb which is used to adsorb the final traces of gas during manufacture of the globe.

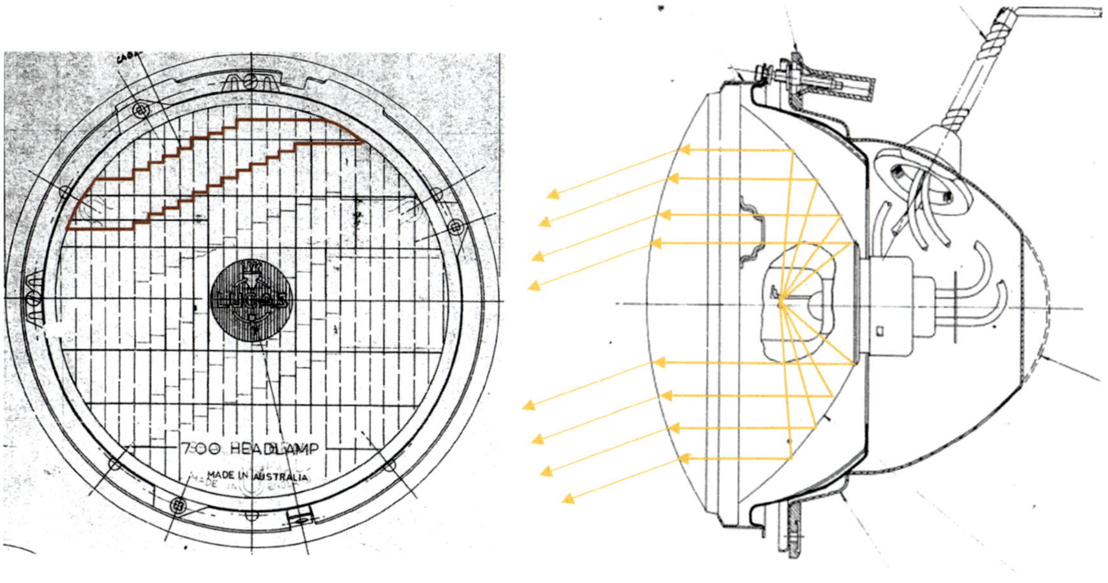

Fig. 12.3.1 *Rays emanating from a headlamp.*

In practice, there are usually two filaments side by side in the globe. The main or high beam filament is located at the focus, while the dipped beam is mounted a little off focus so as to cause the emanating rays to be directed downwards a little rather than straight ahead as shown in Fig. 12.3.1. However, rays arising from the off-focus position are divergent from the outer edges of the parabolic reflector and so must the steered downwards by the glass lens.

The glass lens comprises several different areas which are designed to direct light rays in the desired direction. Fig. 12.3.1 shows one area at the top left where the thickness of glass is greater compared to the region below it. The thickness is calculated so that the rays are projected downwards. These portions of the lens also act on the main beam bringing the light rays downward where they are of more use.

Horizontal spread of the beam is controlled by the vertical flutes in the glass. These act as concave lens which cause the rays to diverge as shown in Fig. 12.2.7. In addition, the divergence is biased towards the left for RHD vehicles and to the right for LHD vehicles.

12.4 Interior Mirrors

There are 32 different interior mirrors used throughout the BMC/Leyland range counting both UK and Australian manufacture. For the Australian Morris Mini range, later models use interior mirrors are common to the Leyland P76 and Marina models.

A review of the Service Parts Lists and drawings for Mini and Moke show the following details:

Parts Book	Model	Mirror	Bracket	Parts Book Picture or Photo
HYL 2980 PUB 1012	ADO15	14A5762 (ND) Clips 14A5763 (14A5578) 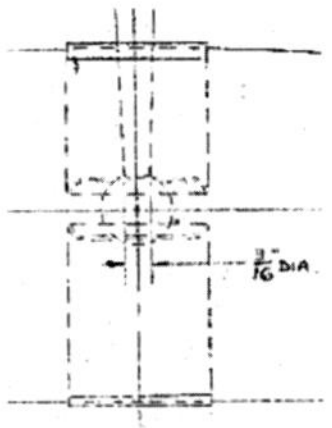 • See discussion at end of Table for more information about clips.	14A7036 (ND) 	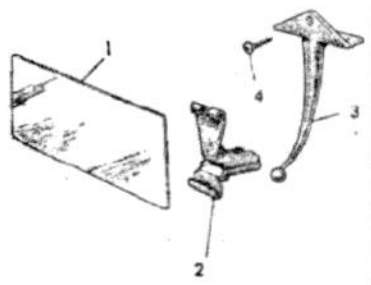
HYL 2980 PUB 1012	ADO15	14G3747 (alternative to 14A5762) 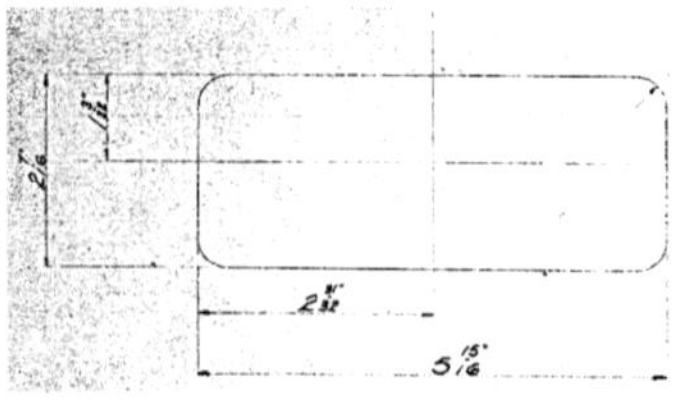	14A7036 (ND) Clips 14A5763 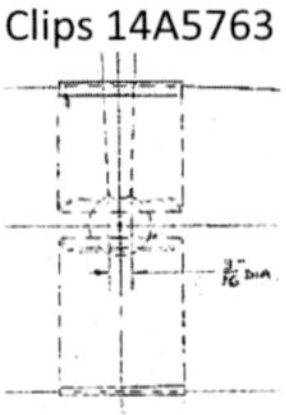 Black crackle finish.	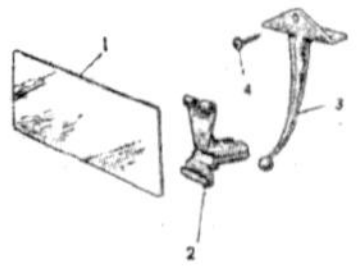

PUB 1056 M	All models	HYA2570 This is an assembly, but Parts Book still lists clips and bracket. 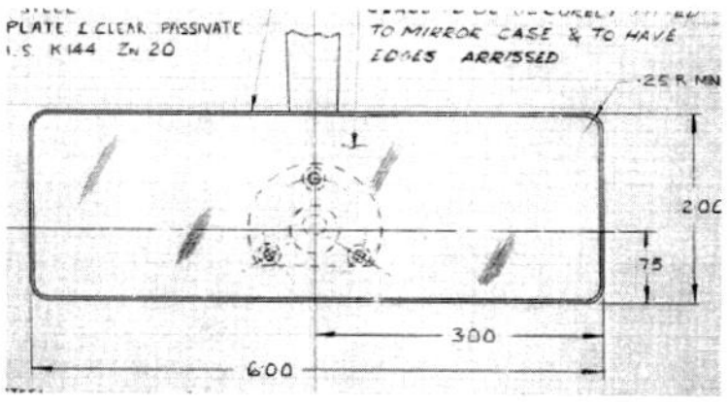 14A5763 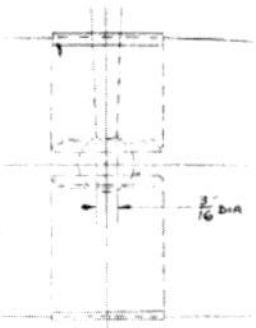 Parts book shows Part No. for straight clips. Picture shows angled clips. Actual assembly has no clips.	14A7036 Bracket in picture looks different and has suction cup, but Part No remains same as previous. Drawing for HYA2570 shows a different bracket and no suction cup. 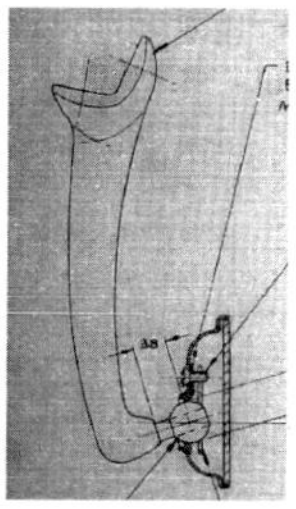 HYA2570	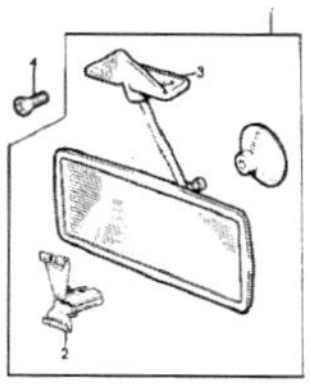
PUB 1052	YDO4 YDO5 YDO6 YDO21 YDO22 YDO23	HYA2570 (Assembly) 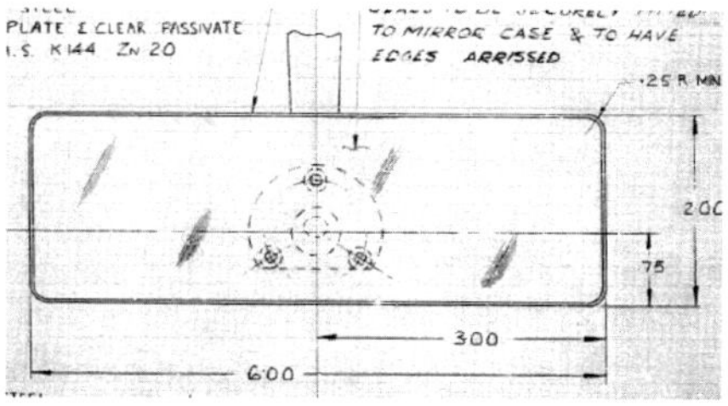	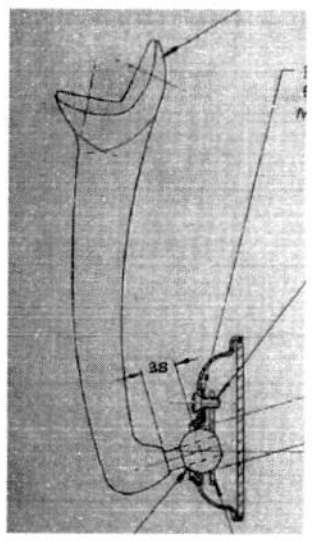	
PUB 23	YDO21 YDO21V YDO22 YDO23 Morris	24A2110 (Assy) (24A1536) 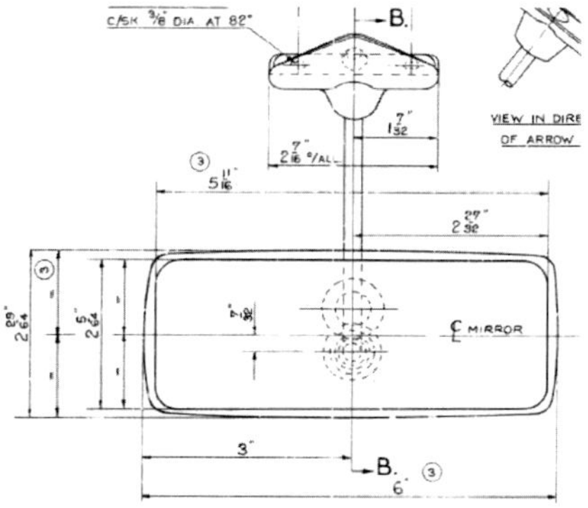	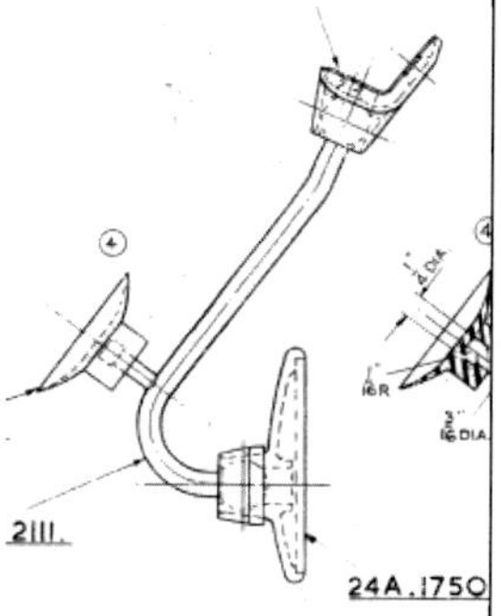	

	YDO21 YDO21V Morris	HYC8229 (HYB3185 Black) (HYC5891 Beige) 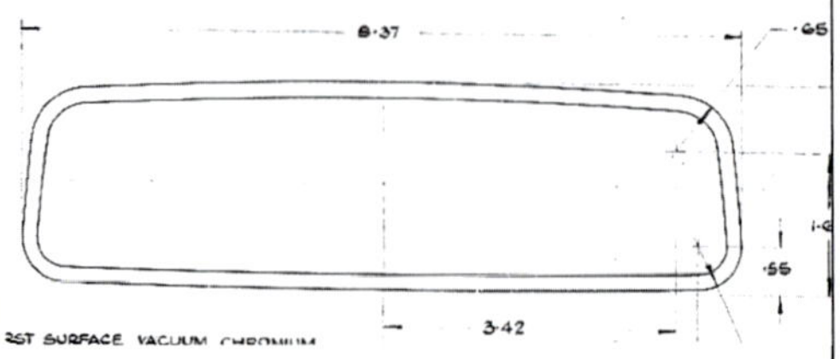 Same as Tasman/Kimberley	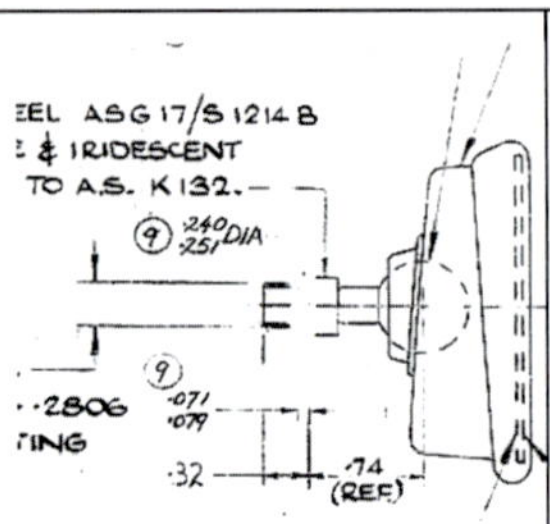	 
	YDO21 YDO21V Leyland	HYC7761 (Black) 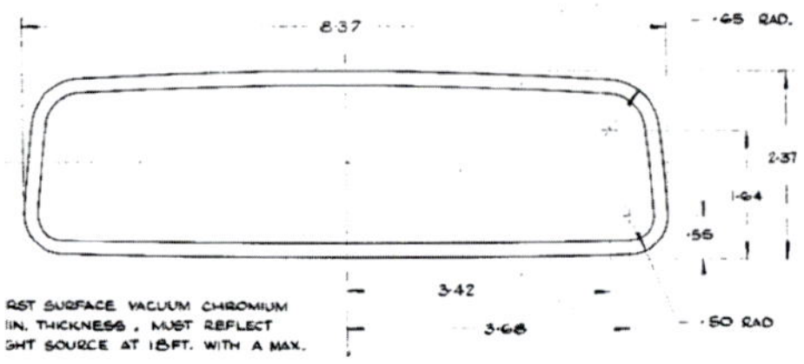 Same as P76. Note, no tilt function.	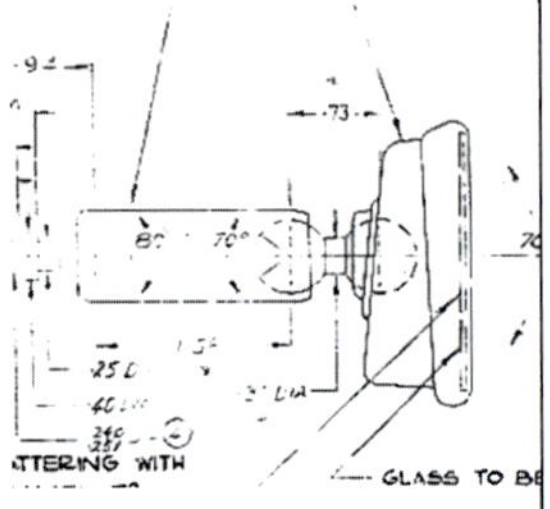	
	YDO22 Leyland S, SS and LS	HYC7797 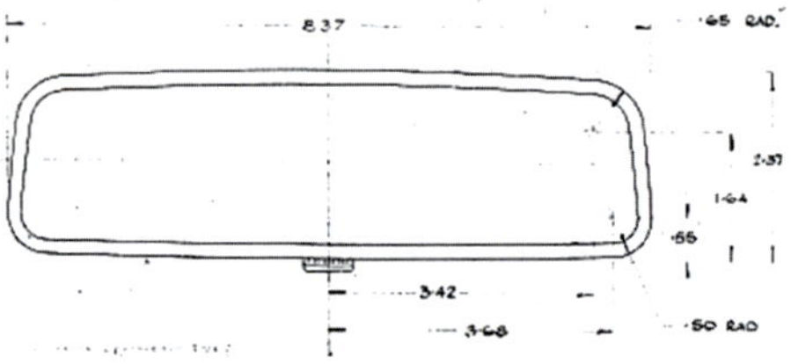 Same as P76.	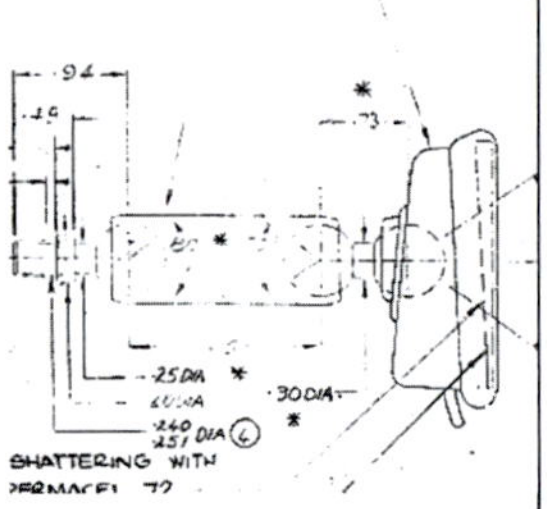	
	YDO21V Morris & Leyland	HYB7427 (Assy) (Grey) (HYD302) 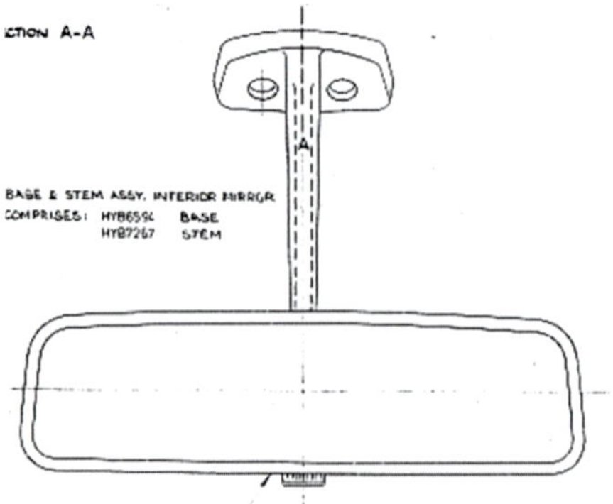 Same as Marina	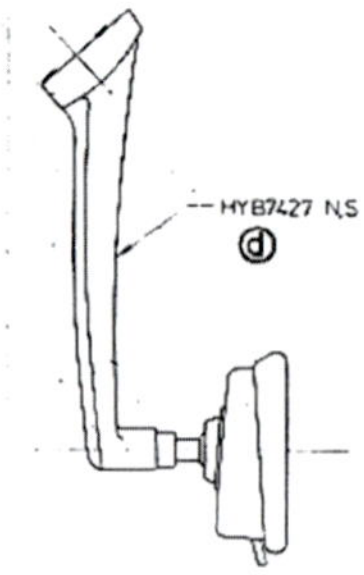	

	YDO21V Morris & Leyland	HYB8145 (Assy) (Black) (HYB4781) 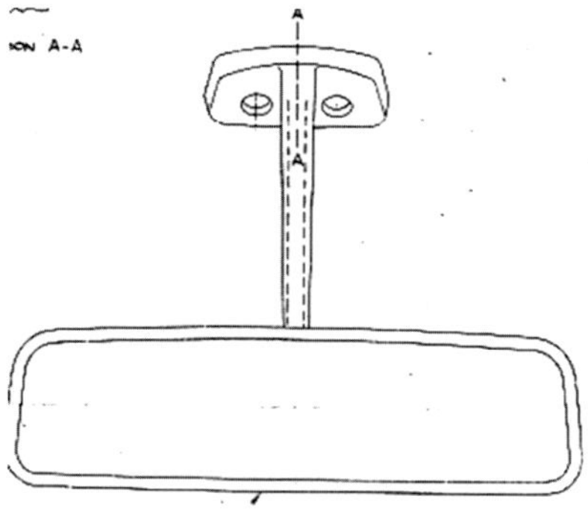 Same as Marina. Note, no tilt function.	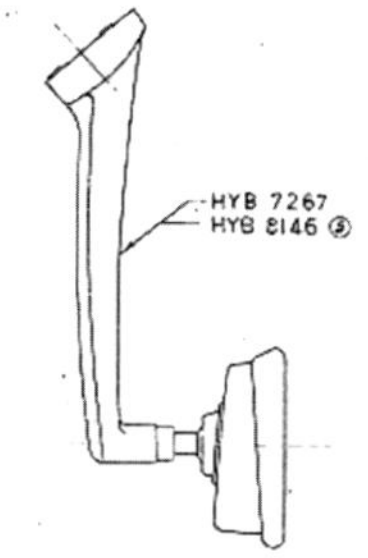	
	YDO21 YDO22 (Morris & Leyland)	HYC7750 (Assy) (Black) Same as Force 7. Parts book shows HYC7750 as an assembly, but the drawing is for the stem only. Stem has breakaway notch in accordance with ADR requirements. 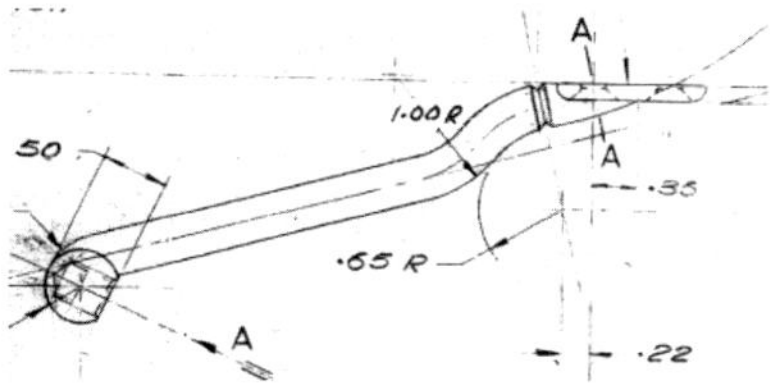	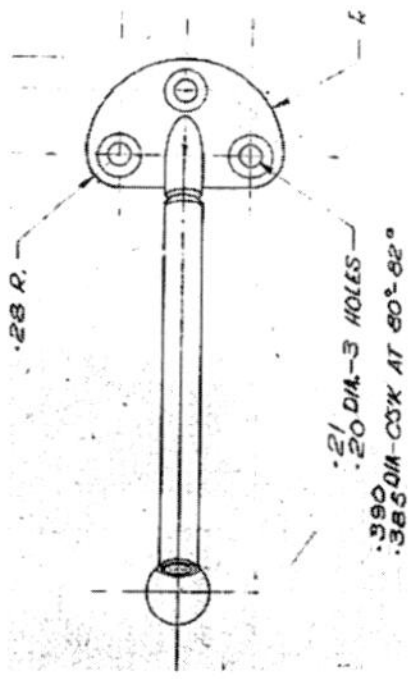	

Table 12.1.1 *Mirror fitment, Mini range.*

ADO16 Morris 1100				
1013 M	ADO16	ADA6212 (Assy) comprising 34G253 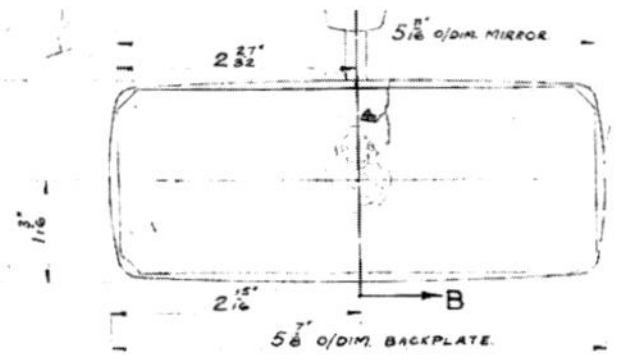	34G254 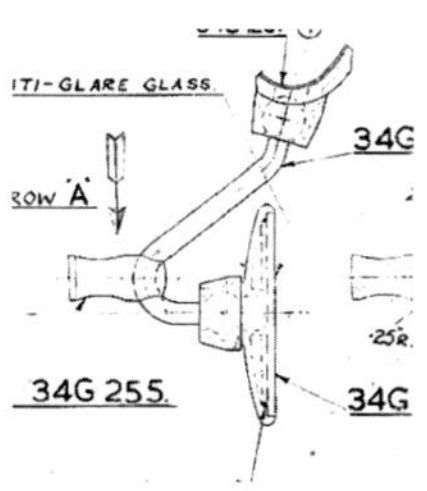	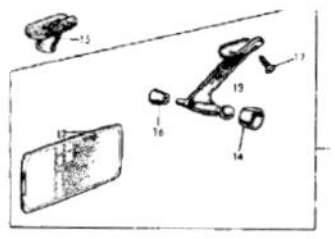
	ADO16	34G1159 (Assy) comprising 24A1533	34G2149	

		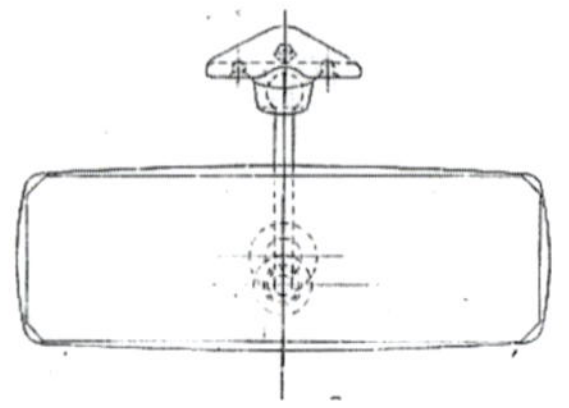	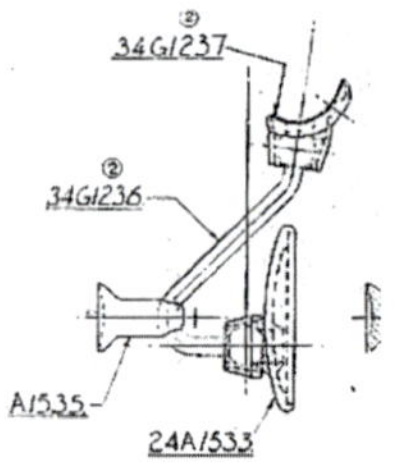	
	ADO16	HYA2486 (Assy) 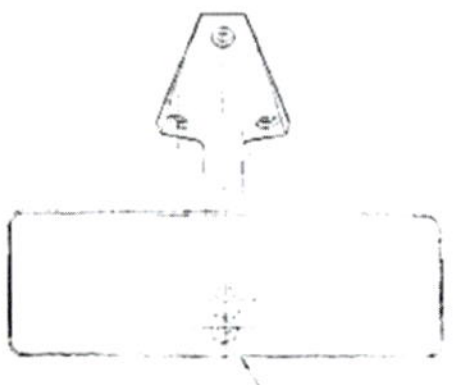	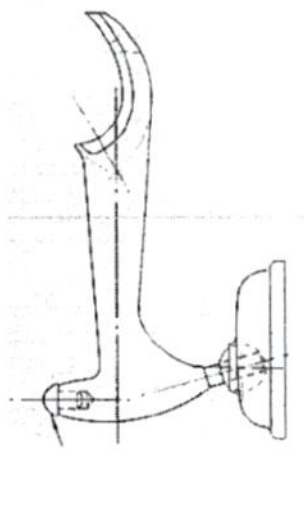	

Table 12.2.1 *Mirror fitment ADO16 Morris 11000 range.*

Stems for early mirrors are marked Eversure Part No M880. There are two styles of clips. Straight clips 14A5763 are shown on Drawing 14A5578. It should be noted that these clips actually come as a pair, one with a different height than the other, to allow for a 180° turning of the mirror for driver height adjustment. Although no drawing can yet be found for angled clips, observation of actual parts shows those two are also of different height.

It is sometimes found that for mirrors with a plastic ball end, things have deteriorated to the extent that no more adjustment can be made to the three securing screws. The ball end can be replaced, but with some difficulty.

Fig. 12.3.2 *Replacement Delrin ball end and fitment to mirror stem and body.*

The picture above shows the assembly under the glass. The adjustment plate is removed, the old ball end discarded, and a new ball end fitted. The ball end is a ½" diameter Delrin ball with a ¼" hole. For fitment, the stem of the mirror mount is heated with a large soldering iron, and then the new ball pressed on. The adjustment plate is then re-fitted. Although fairly straight forward in principle, the difficulty is that the glass has to be removed first. The backing plate has also to be put into position before the ball end is pressed on.

Appendix

What Thread is That?

Traditional threads used in English machinery are British Standard Fine (BSF), British Standard Whitworth (BSW) and the British Association (BA) threads. Thread pitch is measured in "threads per inch" (TPI) and the included angle between the faces of the threads is 55°.

Spanner sizes are given as the diameter of the bolt – so a ½ Whitworth spanner is a fairly large spanner designed to fit across the flats of very large hexagon head of a ½ diameter bolt or nut. The BA threads are usually found in smaller electrical parts and has a finer thread than the equivalent BSW or BSF diameters.

During World War II, a unified thread system was developed under the auspices of the Society of Automotive Engineers (SAE) in the United States which gave rise to Unified National Coarse (UNC), and Unified National Fine (UNF) threads. In these systems, the pitch is still measured in TPI, but the included angle is 60°. Spanner sizes are given as the dimension across the flats of the bolt head (A/F). So, a ½ SAE spanner has ½" gap between its jaws. The actual diameter of the bolt is much smaller (about 5/16).

	UNC			**UNF**			
Diameter (inches)	tpi	Tapping drill (mm)	Clearance drill (mm)	tpi	Tapping drill (mm)	Clearance drill (mm)	Spanner size (A/F)
1/4	20	5.1	6.5	28	5.5	6.5	7/16
5/16	18	6.6	8.1	24	6.9	8.1	1/2
3/8	16	8	9.7	24	8.5	9.7	9/16
7/16	14	9.4	11.3	20	9.9	11.3	5/8
1/2	13	10.8	13	20	11.5	13	3/4
9/16	12	12.2	14.5	18	12.9	14.5	13/16
5/8	11	13.5	16.25	18	14.5	16.25	15/16
3/4	10	16.5	19.25	16	17.5	19.25	1 1/8
7/8	9	19.5	22.5	14	20.4	22.5	1 5/16
1	8	22.25	25.75	12	23.35	25.75	1 1/2

Table A1. *Comparison of thread sizes.*

Smaller sizes than ¼ UNC or UNF are designated by numbers. For example, the next size down from ¼ UNC is a No. 12 thread, often written 12-24 which means No. 12 thread, 24 TPI. An equivalent diameter UNF size is 12-28, indicating 28 TPI.

Notice that the tapping drill size for a UNF threaded hole is larger for a UNF thread compared to UNC – that is, the coarse threads cut deeper, and have more spacing between them than the fine threads, but the fine threads have a shallower angle, and so a greater "wedging" mechanical advantage when being tightened.

Although the spanner sizes are different, a BSW threaded bolt will screw into a UNC threaded hole, but this should never be considered acceptable since the included angles are different – and so the frictional and load bearing characteristics of the joint will be compromised.

In BMC vehicles, most nuts and bolts are UNF where steel parts are joined – a fine thread. Coarse UNC threads are used when the bolt is to be screwed into an alloy part such as the clutch housing. Some threads (e.g. the starter motor cable post) are Whitworth as is the gearbox mainshaft nut (LH on Austin 1800, RH on Mini). Some threads in the transmission are designated UNS. These special threads have a finer pitch than the UNF series and are designed for high stress applications.

Other books of interest:

	**Leyland Cars in Australia: a Chronicle** ISBN 978-0-9585525-5-4, 197 A4 pages. This book tells the story of The British Motor Corporation, Leyland Australia, and Jaguar Rover Australia from its inception in the 1950's to its demise in the late 1990's. A detailed history of the culture, the plants, the people, and the vehicles.
	**BMC-Leyland Australia Reference** Peter J Davis and Tony Cripps ISBN 978-0-9585525-6-1, 258 A4 pages. This book provides extensive details of vehicle identification codes and power unit fitment to Australian-made BMC/Leyland passenger cars and derivatives from 1950 to 1982. Crammed with vehicle identification information and cross references. Your bible of BMC cars.
	**Engineering Series** **Morris 850** ISBN 978-0-9585525-7-8 **Morris Mini Deluxe** ISBN 978-0-9585525-8-5 **Morris Cooper S** ISBN 978-0-9585525-9-2 **Mini Clubman & GT** ISBN 978-0-6452067-1-5 **Leyland Mini** ISBN 978-0-6452067-5-3 **Morris Mini Moke** ISBN 978-0-6452067-3-9 **Leyland Moke** ISBN 978-0-6452067-7-7 **Austin 1800** ISBN 978-0-6452067-9-1 Each of these unique books presents excerpts from over 500 factory engineering drawings for each of these vehicles.
	**Secrets of Style** David Bentley, David Hardy, John Holt and Tony Cripps. ISBN: 978-0-6452067-2-2, 72 A4 pages. More than a historical record about vehicle styling at BMC, this book educates the reader's eye so that you may look upon both modern and classic vehicles with a new understanding and appreciation of the art of motor vehicle styling
	**It's a Mini World.** Tony Cripps, Peter Davis, Bruce Elson, Will Hagon, Nairn Hindhaugh, & Ron Moss. ISBN 978-0-6452067-0-8 A new book which commemorates the production of classic Mini and Moke in Australia. Written by those who worked at the factory. Anecdotes, production methods, advertising, servicing, identification, and a great line up of appendices.

www.leylandaustralia.com.au

242 THE BMC ENGINEERING COMPANION

Made in United States
Orlando, FL
04 November 2026

90200367R00145